P9-DVQ-895

BETTER HEALTHCARE THROUGH MATH

SANJEEV AGRAWAL & MOHAN GIRIDHARADAS

BETTER HEALTHCARE THROUGH MATH

BENDING THE ACCESS AND COST CURVES

ForbesBooks

Published by ForbesBooks, Charleston, South Carolina.
Member of Advantage Media Group.

ForbesBooks is a registered trademark, and the ForbesBooks colophon is a trademark of Forbes Media, LLC.

Printed in the United States of America.

10 9 8 7 6 5 4 3 2

ISBN: 978-1-95086-334-1
LCCN: 2020914489

Cover and layout by David Taylor.

Advantage Media Group is proud to be a part of the Tree Neutral® program. Tree Neutral offsets the number of trees consumed in the production and printing of this book by taking proactive steps such as planting trees in direct proportion to the number of trees used to print books. To learn more about Tree Neutral, please visit **www.treeneutral.com**.

Since 1917, Forbes has remained steadfast in its mission to serve as the defining voice of entrepreneurial capitalism. ForbesBooks, launched in 2016 through a partnership with Advantage Media Group, furthers that aim by helping business and thought leaders bring their stories, passion, and knowledge to the forefront in custom books. Opinions expressed by ForbesBooks authors are their own. To be considered for publication, please visit **www.forbesbooks.com**.

Mohan: *To my wife Noona, and our daughters Shirali and Prianka, for hanging in there through the roller-coaster ride of a startup, and to my incredible colleagues at LeanTaaS who made the roller-coaster ride so exhilarating.*

* * *

Sanjeev: *To my family—my daughters Shalini and Sonia, who are daily reminders that working to improve healthcare is worth it, and my wife Bhanu who has supported all my crazy endeavors—and to my LeanTaaS team that has been 100 percent committed with head and heart to our mission.*

CONTENTS

FOREWORD

Allan D. Kirk, MD, PhD, FACS

Is healthcare a business or a calling, a right or a privilege? Do we care for patients or customers? Do we rely on market forces or communal safety nets? Should we strive for efficiency or resilience?

The answer is … yes, which does not seem like a generalizable answer.

The problem is not the answers; it is the form of the questions—this *or* that. The questions, as posed, are nonsensical. How we *care* for one another (it is, after all, health*care*) is not dichotomous. People are decidedly not digital; they are wonderfully analog, and despite our growing love for ones and zeros, there are far more than two sides to the equations defining their lives. Hard problems, like healthcare, are multidimensional, not binomial. Thus, an approach to the pervasive problems in healthcare—cost, quality, access—has to encompass not only numerous variables, but also the ever-expanding and ever-changing sets of variables that uniquely define a patient's situation within a system, and the system's situation over time.

Unfortunately, the common approaches to healthcare have

been almost exclusively locked in simplistic, fixed models defined by deterministic variables: x number of patients divided by y number of doctors equals a schedule. Solutions have been expressed in rigid, artificially uniform outputs: you have twenty patients today booked into equally divided fifteen-minute time slots. Intuitively, we all know this doesn't work. Patient problems are stochastic; there is no "average patient." No effective appointment starts and ends by the clock, but rather when the problem is solved, and this requires a probabilistic rather than deterministic approach. Importantly, static models become useless when things change, which they always do (particularly in 2020). Solutions need to be dynamic, flexible, and, ideally, become more effective with growing experience.

Enter artificial intelligence (AI). This is not new, nor are its progeny, machine learning and neural networks. They have been around for decades, but their practical application has been markedly enhanced by exponential increases in computing speed and power, and in the last decade, AI has been influencing your life more than you think. The math underlying AI is not all that complicated; it is related to statistical regression analysis in many ways. However, the ability for AI algorithms to deal with missing information and uncertainty and still perform beyond the bounds of the programmer separates it from statistics. AI is why search engines work, why you can find exactly what you were looking for on Amazon, and why you can find an Uber to take you almost anywhere, any time.

If consumer businesses have mastered this, why is it taking so long for healthcare to get here? It is certainly not because people in healthcare are not smart or are disinterested; it is because the stakes in healthcare are infinitely higher. It has been important to make sure we understand how AI works when all that is at stake is getting a coffeemaker delivered—before trying it out in a busy emergency room or when scheduling someone's open-heart surgery. Furthermore, while analogies to the airline industry and Uber are OK, they only go so far. The complexity and importance of healthcare quickly outpace consumer applications. Nobody dies if your plane doesn't take off on time, while a few minutes can mean life and death in the operating room. Furthermore, we understand planes (we built them)—that is not the case for the people. So rather than fixating on why AI has not been embraced sooner, it is best to view its emergence in healthcare as an approach that is now graduating to the big leagues, the league of life and death—as opposed to convenience and luxury.

This is not a math book. Indeed, the word *equation* only shows up one time! It is also not a primer of AI. What it is, is a book introducing the power of AI and algorithmic learning to the common problems plaguing healthcare delivery. It is conceptual and exemplary. It introduces the way healthcare must go, and why.

It is certainly timely that the authors have put this book out now, as the need to simultaneously solve the healthcare equation for efficiency and resilience has never been more evident than in the year of its publication. My health system adopted an algorithmic approach to operating-room management last year. We used it to make it easier to get patients to their surgical teams with record efficiency, which was good. However, without changing anything, we then used the same process to respond to the challenges of the global coronavirus pandemic, something nobody anticipated. By building an operational model that can function at any volume, fluidly solving for efficiency as we vary the myriad inputs of patient complexity and environmental variation, we were able to maximize our care when going full speed, and judiciously manage our care (protecting our patients and our staff) when the brakes were unexpectedly slammed on. Indeed, the tools outlined in this book helped us deliver the care

PREFACE

EFFICIENCY IS NO LONGER A LUXURY

Amy has just been told she needs to have a total knee replacement. After getting a second opinion from a primary care provider, she starts doing some online research—and talking with friends and family—to identify the "best" few orthopedic surgeons in the community who could perform the procedure. Luckily, Amy has excellent health insurance; she confidently sets about trying to secure an appointment to get the process started.

If Amy is lucky, she will be given an appointment date for the first consultation that is five to thirty days out. If she is not as lucky, her appointment might be scheduled as far as two months or more down the road. Once she sees her physician and finds out she needs that knee replacement, the scheduling of that procedure may take another few weeks. Meanwhile, all that she, and her physician, can do is wait. And hope. And *wonder*—why is it that a service appointment for her car can be scheduled within the same week, yet when an important health matter is involved, she is forced to play the waiting game?

The above scenario repeats itself hundreds of thousands of times a day in homes and medical centers across the country. Long lags in appointment times are so commonplace that they are considered the norm, not an aberration. Yet this state of affairs is a lose-lose-lose proposition for everyone involved. No one is doing the happy dance here—not the patient, not the physician, not the medical facility.

EVERYONE WANTS IMPROVED EFFICIENCY, BUT ...

Patients naturally want their interventions performed in as timely a manner as possible. They are often feeling some anxiety about their health—that's why they went to the physician in the first place. Every day they are forced to wait is an added day of tension and worry. If they are in pain or discomfort, they may now need to endure this symptom for weeks instead of hours or days. And it's possible their condition may worsen during the waiting period. At the very least, the patient will feel their health concerns were not a priority for the provider or the system.

> When you try to cram ten pounds of tomatoes into a five-pound bag, something needs to give. And the result usually isn't pleasant.

Physicians and other healthcare providers want to provide the best care possible for their patients. They are frustrated when the resources they need in order to *provide* that care are hard to access. So, when they are unable to schedule a procedure in an efficient manner, they often shoehorn the appointment into a day that is already fully booked. This results in jammed schedules, which creates a mess for everyone. Waiting rooms become crowded, patients sit for hours, support staff are overburdened and end up working overtime. When you try to cram ten pounds of tomatoes into a five-pound bag, something needs to give. And the result usually isn't pleasant.

The medical facility, for its part, wants to offer good customer service. It knows that when patients are assigned lengthy waits for appointments, they may opt to go elsewhere for their care (provided the facility/insurance plans permit such choice). When this happens, the facility runs the danger of losing the patient forever. Patients are not just one-off customers, after all; they have a "lifetime value." A patient who comes in for an oncology consult, for example, might return for radiation treatments, infusions, surgical procedures, and more over the ensuing months and years. Losing a patient can mean losing an entire income stream.

Lose-lose-lose. No one likes the current system.

WHERE DOES THE PROBLEM LIE? SUPPLY AND DEMAND

Yet here's the baffling part. There are probably ten opportunities *every single day* between the date the appointment is made and the date the procedure is scheduled when the patient could have been slipped onto the calendar. How do we know?

Let's say the appointment involves a sophisticated imaging machine. If you were to stand next to that machine all day, every day, you would probably observe several periods each day when the machine sat idle for twenty to thirty minutes without a patient inside it being actively scanned. (Don't believe us? If you have access to one of these machines, check its daily time stamps.)

So why wasn't the patient able to secure an appointment for

the next day or at least within the same week? Let us give you a few hints…

- It wasn't because the staff member didn't care about the patient's health or convenience. Laziness and indifference are not the culprits here.
- It wasn't because the facility was holding appointment slots for higher-priority patients. The system isn't rigged.
- It wasn't because the system was incapable of displaying all the possible appointment slots. Virtually every medical office uses an electronic health record (EHR). The problem isn't an IT issue.

The reason for the systemic inefficiency comes down to the problem of matching supply and demand for every single appointment. Turns out, it is very difficult to align an unpredictable pattern of demand for appointments with a limited supply of physicians, staff, equipment, and rooms—a supply that is further restricted by a host of real-world constraints.

On the demand side of the equation, it is very hard to predict how many patients a clinician or a surgeon will be seeing on a given day, how complex those individual cases will be, and what procedure or appointment might need to be scheduled as a follow-up. Patients don't get sick on a schedule. Some days a given physician may see as many as, say, twenty-one patients; other days the same physician may see as few as nine. So, although the average patient load may be fifteen, that tells us nothing about how many patients will be coming in *on any specific day in the future*—let alone how many people will walk into an ER on any given day, within any given hour.

Meanwhile, nurses, rooms, receptionists, and equipment are in fixed supply on any given day. Matching varying demand with fixed supply is always tricky. And then there are the added availability constraints. Physicians often split their time across multiple locations (e.g., their clinic and the hospital where they do their rounds or procedures). Certain kinds of nurses are required to perform certain procedures. Midday hours bring in a flood of patients, all needing the same rooms or the same pieces of equipment, and so on.

Every medical facility employs three major asset types—people (physicians, nurses, technicians), equipment/supplies (machines, pumps, robots, drugs, implants), and facilities (rooms, chairs, etc.).

In order to execute a healthcare appointment, all of the required assets must line up *and be available at exactly the same time.*

In order to execute a healthcare appointment, all of the required assets must line up *and be available at exactly the same time.* For a chemotherapy treatment to occur, for example, the nurse, the pump, the drugs, and the chair must all be available at the time in question. If even one asset is missing or unavailable, the appointment cannot take place. And each of these assets has its own scheduling anomalies—people call in sick, equipment breaks down ...

So you can see why matching supply with demand in a clinical environment isn't as easy as it sounds. To top things off, the tool most facilities use to orchestrate this matchmaking feat is essentially a calendar. Nowadays a calendar might be displayed on a computer and contain a host of "smart" features, but a calendar is still a calendar.

MONEY IS A MAJOR FACTOR TOO

Supply and demand wouldn't be a big problem if the assets in question were not in scarce supply. Some of the equipment needed can cost as much as ten, twenty, or even one hundred million dollars. These are expensive resources that require a lot of planning prior to being deployed—you can't just casually place an order to buy more of them to meet increasing demand. Also, the personnel involved are highly trained physicians, nurses, and technicians who spend years completing their training. It is very difficult to just "go out and hire some more people." You must utilize your current personnel and equipment as efficiently as possible—especially given the financial challenges the healthcare industry is currently facing and will be facing more acutely in the coming decades.

We'll talk about some of those challenges in a moment, but for now let's just note that the United States has the highest per-capita healthcare costs of any nation in the world. As of this writing, healthcare devours 18 percent of our GDP, with that percentage expected to rise to 20 percent within the next few years.[7] Meanwhile, reimbursements for services are going steadily downward for healthcare providers.[8] Bottom line: there isn't going to be any "extra" money in the system. Quite the contrary. More will need to be done with less. All of this adds up to a scenario in which efficiency is no longer a "nice to have," but rather a "must have."

---

7 John Commins, "Healthcare Spending at 20% of GDP? That's an Economy-Wide Problem," HealthLeadersMedia.com, September 19, 2018, https://www.healthleadersmedia.com/finance/healthcare-spending-20-gdp-thats-economy-wide-problem.

8 Robert Fifer, "Health Care Economics: The Real Source of Reimbursement Problems," The American Speech-Language-Hearing Association, July 2016, https://www.asha.org/Articles/Health-Care-Economics-The-Real-Source-of-Reimbursement-Problems/.

TIME FOR A PARADIGM SHIFT

Yet when it comes to day-to-day scheduling inefficiencies, most medical centers are trapped in a mindset of "this is just the way things are." Operationally, many of them are still in the last century. It is odd, isn't it, that in a healthcare industry that excels in both medical and technological advances, we freely accept a glorified pencil-and-paper method for managing our precious assets?

And make no mistake, the United States *is* the global leader in medicine, despite its system's many flaws and challenges. Most surveys still rank the United States as number one in the quality of its physicians. Four of the world's top ten hospitals, according to *Newsweek*, are in the United States, while fourteen of the thirty most technologically advanced hospitals are US-based.[9] More than half of the world's new medicines are developed in the United States,[10] and five of the top ten biotech companies are located here.[11]

In our opinion, this country's advances in clinical excellence over the past few decades have been nothing short of spectacular. However, the US healthcare system has *not* made a similar investment in *operational excellence*. For that reason, it has struggled to keep up with the ever-increasing complexity of providing clinical care in today's world.

---

9 "30 Most Technologically Advanced Hospitals in the World," TopMastersin-Healthcare.com, March 24, 2014, https://www.topmastersinhealthcare.com/30-most-technologically-advanced-hospitals-in-the-world.

10 Grace-Marie Turner, "Though the US Is Healthcare's World Leader, Its Innovative Culture Is Threatened," *Forbes*, May 23, 2012, https://www.forbes.com/sites/gracemarieturner/2012/05/23/though-the-u-s-is-healthcares-world-leader-its-innovative-culture-is-threatened/#2ae0da4977eb.

11 Sintia Radu, "US, China Compete for Medical Research Leadership," USNews.com, September 27, 2019, https://www.usnews.com/news/best-countries/articles/2019-09-27/china-threatens-the-us-leadership-position-in-medical-research.

This reluctance to invest in operational excellence has resulted in a strange paradox: "We're so busy, it's impossible to find a timely appointment slot for our patients, yet many of our valuable assets are sitting idle for several hours each day."

And it's not as if there aren't successful models of efficiency toward which healthcare can look. Examples abound of industries and businesses that have been using the power of math to solve mind-bogglingly complex issues of scheduling, asset utilization, and supply and demand.

- UPS and FedEx guarantee delivery the next morning by 10:30 a.m. anywhere in the country, despite the fact that they have no way of knowing how many packages will be shipped from any origin to any other destination on any given day.

- Amazon knows what a user is going to buy almost before the user does. Using predictive models and an increasingly precise set of analytics that can classify a user based on multiple characteristics, Amazon enables businesses to target the right user at the right time with the right product and offer—not just online but increasingly in the physical world too.

- Uber deploys almost four million drivers worldwide and manages fourteen million daily rides.[12] Its operations system must coordinate geolocation, live messaging, booking, billing, payments, push notifications, price calculations, ratings, tipping, and more for each of these drivers and passengers, in real time. It must also ensure there is always an adequate supply of drivers to meet rider demand in a world where weather, concerts, sporting events, and other factors

12 Mansoor Iqbal, "Uber Revenue and Usage Statistics (2020)," updated July 14, 2020, https://www.businessofapps.com/data/uber-statistics/.

can shift demand in an instant.

- OpenTable has given restaurants the ability to fill seats in a way they never could in a world where phone calls and emails were the methods by which diners made reservations. This has led to massive improvements in restaurant efficiency and in consumer happiness with the booking process.
- Airlines have done a remarkable job of using data science and prescriptive analytics to turn around a historically inefficient industry. In 1999 the airlines were a $100 billion industry, yet all of the airlines (except Southwest) were going through boom-and-bust cycles, and many of them filed for bankruptcy at least once. They were hurting financially. By 2019, the industry had consolidated while doubling in size. Its ticket prices had risen at a lower rate than inflation, yet its profit margins had soared from negative territory to +9 percent. How did the airlines accomplish this turnaround? By learning to precisely match uncertain demand with a fixed, expensive supply of assets, route by route and day by day.

All of these companies are living proof that achieving true digital transformation is possible. But in order to accomplish this, these companies needed to reimagine their core processes. They did this by using advanced math, sophisticated algorithms, and cloud-based software deployed at full scale.

So how can healthcare do the same?

That's what this book is about.

MEET THE AUTHORS

We, the authors, Sanjeev and Mohan, are senior executives at LeanTaaS, a software company that focuses on improving healthcare operations.

Over the past six or seven years, we (and our team of several hundred talented professionals) have conducted thousands of detailed conversations with physicians, nurses, administrators, and healthcare executives to understand the issues they face and to sharpen our perspectives. Our company has invested well over $100 million in building software products that address these problems by mathematically matching supply and demand for healthcare assets. We have helped over a hundred leading healthcare systems in the United States realize substantial improvement in their operational performance. As a result, millions of patients have enjoyed improved access to care and shorter wait times. Health systems have been able to recruit more surgeons, despite feeling they could not guarantee them permanent block time. Providers and other staff members have experienced a marked reduction in the chaos they had come to accept as their day-to-day reality. In other words, we have helped create win-win-win situations for patients, care providers, and operations teams alike.

We like the win-win-win outcome much better than lose-lose-lose—and so do our health system customers.

It is possible to *simultaneously* improve patient access, reduce costs, and shorten patient waiting times.

This book is intended to share the lessons we have learned in building scalable software products that have transformed the operational performance of the many health systems we have worked with. As the pages unfold, we will make some provocative assertions that will challenge the conventional wisdom you may be accepting as true. We won't delve heavily into the math itself; rather, we will explain our logic using analogies from other industries. These examples will make immediate sense to you based on your own experiences. Ultimately, we will lay out a road map that will help you determine the right path for you as you seek to optimize operations within your health system.

WHO CAN BENEFIT FROM READING THIS BOOK

This book will be of keen interest to the following people:

- Leaders in healthcare provider organizations (C-suite, VPs of cancer, surgery, ambulatory, etc.)
- Healthcare operational leaders (heads of oncology, specialty clinics, inpatient beds; VPs of performance excellence, etc.)
- Payers and policymakers aspiring to increase access while lowering the cost of healthcare
- Healthcare technology vendors who have experienced the operational challenges in healthcare and may be interested in a distinctively different approach to solving the problem
- Anyone—students, entrepreneurs, investors—interested in making a contribution toward improving healthcare in the United States

***

Healthcare is in a bind much like the one airlines were in years ago, when deregulation and competitive pricing were forcing them to become more efficient. Healthcare providers *must* become more muscular in their operations if they wish to deliver the high-quality experience that their patients are increasingly demanding.

Fortunately, there is a solution for many of their challenges. All that is required is a fresh mindset.

And some good math.

Let us show you how we do it.

PART ONE

CHAPTER 1

THE LOOMING CHALLENGE

If you work in the healthcare industry, or even if you're just an interested observer, you don't need a book to tell you that the financial pressure is on as never before. A perfect storm of circumstances is swirling together, one that will make *survivability*, not to mention *profitability*, a greater challenge for healthcare companies than we've seen in the modern era.

As with banks, retailers, and airlines, which had to rapidly enhance their brick-and-mortar footprints with robust online business models—it is the early movers eager to gain new efficiencies that will thrive and gain market share. The slow-to-move and the inefficient will end up being consolidated into larger health systems seeking to expand their geographical footprints.

THE PRESSURES ON HEALTHCARE

Let's look at just a few of the looming challenges healthcare must meet head-on.

AN AGING POPULATION

By the year 2030, the number of adults sixty-five years of age or older will exceed the number of children eighteen years or younger in the United States. We are living longer lives than our parents did. Positive news, for sure, but also problematic for several reasons.

The older we get, the more medical help we need. Older people have more chronic disease. By 2025, nearly 50 percent of the population will suffer from one or more chronic diseases[7] that will require ongoing medical intervention. This combination of an aging population and an increase in chronic disease will create a ballooning demand for healthcare services.

Furthermore, as the population *in general* continues to increase every year, both young and old people will be added disproportionately to the mix, and these are the two segments of the population that need the most medical care. So not only will there be more life spans to take care of but also, on a per-life-span basis, we will need more care than we did before.

LESS MONEY IN THE POOL

Now throw in the economics. The healthcare system as a whole is under tremendous financial strain. As previously mentioned, US healthcare costs per capita are already the highest in the world and rising. Health insurance premiums have increased 54 percent in the past ten years, with Americans over age fifty-five paying even greater increases.[8] The cost of healthcare is the single greatest financial

---

7 "The Growing Crisis of Chronic Disease in the United States," Partnership to Fight Chronic Disease, accessed July 2020, https://www.fightchronicdisease.org/sites/default/files/docs/GrowingCrisisofChronicDiseaseintheUSfactsheet_81009.pdf.

8 Alicia Adamczyk, "Health Insurance Premiums Increased More Than Wages This Year," CNBC Make It, updated September 26, 2019, https://www.cnbc.com/2019/09/26/health-insurance-premiums-increased-more-than-wages-this-year.html.

problem facing US families today.[9] And as the number of older Americans outstrips the number of younger Americans, the burden for paying for the system will fall increasingly on the smaller number of young people. The math does not look rosy.

Given these conditions, it is wildly optimistic to expect that there will be any new money in the system overall. In short, the healthcare system will need to make do with its current pool of dollars—adjusted for inflation—or less. That means there will be greater demand on every asset in the medical field, including human assets.

SHRINKING HUMAN RESOURCES

The US healthcare system is currently facing a shortage of physicians and nurses. In many parts of the country, particularly rural areas, that shortfall has already reached the acute stage. The AAMC (Association of American Medical Colleges) projects that this trend will continue into the foreseeable future and that by 2032, there will be 122,000 too few physicians.[10] By that year the shortage of nurses will be well over half a million.[11]

This problem is closely related to the aging issue. As medicine improves and Americans embrace health trends such as quitting smoking, meditating, and losing weight, people are living longer.

---

9 Jeffrey M. Jones, "Healthcare Costs Top Financial Problem for US Families," Gallup, May 30, 2019, https://news.gallup.com/poll/257906/healthcare-costs-top-financial-problem-families.aspx.

10 Stuart Heiser, "New Findings Confirm Predictions on Physician Shortage," Association of American Medical Colleges, April 23, 2019, https://www.aamc.org/news-insights/press-releases/new-findings-confirm-predictions-physician-shortage.

11 David Alemian, "The Nurse and Physician Shortage," *MD Magazine*, August 9, 2016, https://www.mdmag.com/physicians-money-digest/contributor/david-alemian-/2016/08/the-nurse-and-physician-shortage.

And although medical schools are graduating more physicians, they are not keeping pace with the growing demand created by longevity.

LOWER REIMBURSEMENTS

Reimbursement levels for providers, meantime, are on a downward trend as the payer mix shifts to a higher percentage of Medicare/Medicaid patients. Government health programs pay notoriously less for medical services than most private insurers do. So an aging population also translates to a shrinking revenue stream for health systems on a per-capita basis.

There has also been an inexorable push away from a fee-for-service model toward a value-based care model. The basic premise of value-based care is that the provider is reimbursed based on the *quality* of the treatment, not merely the *provision* of the treatment. So the questions payors are beginning to ask are not just, "Did the patient receive the treatment/procedure as billed?" but also, "Was the patient actually cured?" "Were there any after effects?" "Did the patient need to be readmitted?"

Patient satisfaction metrics are now being factored into payment policies. Payors are using surveys such as the HCAHPS to capture patients' feedback on the care they received, and these satisfaction scores now drive a percentage of the payment many providers receive.

As another aspect of value-based care, providers are increasingly being reimbursed for entire *episodes* of care rather than for each individual service rendered. What this means is that, more and more, services are being *bundled*. This is similar to the way a resort like Sandals operates. When you book your vacation at Sandals, your room, drinks, meals, recreational activities, tips, and more are all included. You don't pay separately for each of these services.

An example of an episode of medical care might be a joint

replacement for a knee. In the "old days," the patient—or their insurer—would pay separately on an itemized bill for the initial consultation, the surgical procedure, the medications, the inpatient bed for one or two nights, the postoperative clinic visits, the physiotherapy required after the surgery, and so on. Now all of those services are being bundled for one inclusive fee. This bundled fee covers everything from the first clinic visit until the recovery is complete. The provider now gets paid less for each of the individual services within the bundle and thus must be more efficient at each point along the chain.

Payors are also now paying for shorter and shorter inpatient stays. A maternity patient can't stay in the hospital for five days anymore unless she wants to pay for some of that time out of her own pocket. She must leave in a day or two. That means the provider must do a skillful and efficient job providing all of the bundled services, such as the ob-gyn, the follow-up nursing care, the nutritionist, the lactationist, etc. All the services must be orchestrated and delivered within the one or two days the patient is in residence, and complications must be avoided. The set of services provided remains the same; providers just have a shorter period of time in which to arrange them. And again, the pay for each of the individual pieces has shrunk.

Finally, telemedicine appointments are on the rise—accelerated by the recent COVID-19 crisis (see epilogue for more on COVID-19). An appointment with a clinical specialist that is conducted over a video call is reimbursed at a far lower rate than the traditional, in-person office visit.

RESTRICTED EXPANSION

Along with the above pressures, health systems also face constraints on their ability to throw resources at challenges. Many of today's smaller health systems do not have the financial strength to "grow their way out of the problem" by building new facilities or expanding current ones. Scaling up has obvious advantages. If you compare a hospital that completes five hundred patient visits per day to another that completes five thousand patient visits per day, the latter does not need to be ten times the size or have ten times as many ORs as the former. Efficiencies are gained. As you increase the number of patient visits completed per day for a given service, the unit cost of delivering that service goes down. (Think of Amazon's or Walmart's per-unit costs.) However, if you don't have the financial resources to expand, you must find a way to see more patients within your existing infrastructure.

It is also difficult for small medical centers to purchase certain pieces of equipment. In order to make and sustain a multimillion-dollar investment in a high-end imaging system, for example, you must have substantial financial resources and the service volumes to support it. The point is that if you're small and don't have substantial resources, it's doubly important that you learn to be operationally efficient. If you own one expensive surgical robot and you're using it badly, you might be tempted to buy a second one to keep up with demand. But perhaps a smarter solution would be to find ways to use the first one more efficiently.

Larger health systems are also facing a squeeze. If you look at the operating income and EBITDA margins of health systems, they have been on the decline of late. Hospital operating margins have shrunk since 2015[12] (and at the time of this writing are facing severe pressures from COVID-19). The hospital construction industry

12 "2019 Health System Financial Analysis," Navigant, October 2019, https://guidehouse.com/-/media/www/site/insights/healthcare/2019/guidehousehsfa2019infographic.pdf.

has encountered challenging conditions as a result, with industry revenue dropping an average of 0.5 percent per year between 2015 and 2019.[13]

As the CEOs of these organizations face pressure from their fiduciary stakeholders or governing bodies, they must demonstrate that they're managing whatever resources they currently possess efficiently and productively.

ATTRITION OF HIGH-REVENUE PAYERS

Margins for healthcare providers are now being sapped in other ways. New competitors in the field are luring away some of the most attractive subsegments of the patient population by offering premium or concierge-style services. One Medical, for example, offers a membership-based primary care experience that ensures quicker and easier access to physicians in exchange for a monthly subscription fee. This option can be appealing for the more affluent (better insured) sector of the population.

Here's why this is a problem for providers. Whenever you have a large diversity of appointment types, patients, and payers, a subset of these appointments tends to subsidize many of the others. Some payers, for example, are able to pay $30,000 for a joint replacement procedure—this can effectively offset those patients who come into the practice with Medicare and receive the identical joint replacement procedure at a substantially discounted rate. But now many of the well-insured are "defecting" for better service. They don't want to wait four weeks to see a physician and sit for two hours in a crowded waiting room. To

13 "Hospital Construction Industry in the US—Market Research Report," IBISWorld.com, March 2020, https://www.ibisworld.com/united-states/market-research-reports/hospital-construction-industry/.

them, an extra $1,000 to $3,000 a year to join a concierge service is a price worth paying.

Whenever you "skim the cream off the top" in this way, you blow up the underlying economics. The airlines learned this lesson with the A380, a six-hundred-seat aircraft that many of them were initially eager to purchase. There were many miscalculations in the A380 initiative, but one of them was this: the last thing busy and successful business executives want to do is jump on a long-haul flight with six hundred other people. The boarding process can be unpleasant, the food is perceived as bad, and the overcrowding at customs and baggage claims can cause delays. So, many of the top-paying customers defected and opted to book their long-haul flights on regular three-hundred-seat aircraft.

When this happens, suddenly your six-hundred-seat aircraft is filled with six hundred backpackers purchasing heavily discounted "supersaver" tickets. And the economics of these routes don't stand a chance. The same type of thing is starting to happen with many healthcare providers.

INCREASE IN AMBULATORY CARE

Finally, the migration of a growing number of surgical procedures from an inpatient setting to an ambulatory care setting is further siphoning dollars away from two of the top sources of revenue in a hospital: the hospital-based OR and inpatient beds. In the "old days," only dental and eye surgery, and a handful of other procedures, could be performed in ambulatory settings. That bar is rising every year, with more and more procedures being handled on an outpatient basis. The trend toward ambulatory care is expected to continue to grow at a rate of nearly 5 percent a year for the next several years.[14] In 2018, the ambulatory care market was around $77 billion; it is expected to hit $113 billion by 2026, a massive leap.[15]

There are many underlying causes for this trend—improved medical procedures, patient preference for shorter treatment cycles, better anesthesia options—but lower cost is undoubtedly a main driver.

A joint replacement performed in a hospital might cost $30,000. If done in an ambulatory setting, it might cost only $15,000. It doesn't take a PhD-level economist to understand why insurers and government payers are fans of the ambulatory option.

But this trend, again, will hurt the economics of hospitals. Previously, revenue from the inpatient beds and OR rooms paid indirectly for a lot of services and costs that weren't reimbursed as lucratively. But now, as inpatient care decreases, there will be less and less of that fulsome revenue coming in.

This will affect the way hospitals do business. Many cancer centers, for example, have historically treated initial clinical appointments as loss leaders. The patient might be billed $300 for the visit. But for the hospital, which needs to pay for its prestigious oncologists, surgeons, and top-of-the-line equipment, the initial consult fee does not even cover its costs. Hospitals have been willing to absorb such losses because they know a certain percentage of these patients will end up requiring infusions, radiation treatment, expensive inpatient treatment care, and other ongoing services. As inpatient care dwindles, however, the financial risks go up for the hospital.

***

14 "Ambulatory Care Service Market to Reach USD 113.06 Billion by 2026," GlobeNewswire, August 7, 2019, https://www.globenewswire.com/news-release/2019/08/07/1898487/0/en/Ambulatory-Care-Service-Market-To-Reach-USD-113-06-Billion-By-2026-Reports-And-Data.html.

15 Ibid.

All of these factors are pulling in one direction only: to eat away at traditional revenue streams for healthcare providers. This strongly suggests to us that health systems *need* to be operating at an extremely high level of executional excellence in order to maximize their existing assets on a day-to-day basis.

> **Health systems *need* to be operating at an extremely high level of executional excellence in order to maximize their existing assets on a day-to-day basis.**

It turns out, however, that healthcare is near the bottom of all industries in terms of asset utilization …

WHY HEALTHCARE IS LAGGING IN ASSET UTILIZATION

There are several reasons for this. Perhaps the biggest *structural* reason that healthcare has historically been able to thrive with a lower asset utilization than other industries is that it is a hyperlocal business with extremely high regulatory barriers to entry. The traditional fee-for-service model with low price transparency enabled health systems to set their prices at comfortable levels and continuously invest in a growing set of assets that could be funded through operational cash flows. This model is undergoing rapid change.

In addition, there are operational reasons healthcare has failed to achieve a high asset-turnover ratio as compared to other industries.

LIMITED SCHEDULES

Healthcare delivery has been built around the personal scheduling preferences of the industry's most important human assets—the physicians, surgeons, and nurses. To recruit the top providers, you naturally need to offer them desirable working conditions. As a result, most healthcare delivery operates on a 9:00 a.m. to 5:00 p.m., Monday through Friday schedule. (Patients, however, have the inconvenient habit of getting sick on evenings and weekends.)

Imagine, however, if the pilots for an airline were only willing to work from nine o'clock to five o'clock on weekdays. It would be impossible to create viable routes (e.g., from the United States to Europe or Asia) or to achieve a profitable level of asset utilization for the aircraft in the fleet. For that reason, the entire scheduling philosophy of the airlines is built on the premise of keeping each plane in the sky for as many hours as possible every day. Same thing in manufacturing. If you purchase a quarter-billion-dollar paper machine, you run it around the clock. Three shifts, twenty-four hours a day, seven days a week, 365 days a year, with very tight downtimes—because every minute the machine is down represents thousands of dollars of uncaptured revenue.

By contrast, some of the most expensive equipment in the healthcare industry is unavailable for use literally *two-thirds of the time*. It is active only during regular office hours. So even perfect utilization would represent only 33 percent of its total capacity.

VOLATILE DEMAND

Another factor is that the demand for healthcare services is highly variable and fluctuates substantially from day to day. Understandably, this forces health systems to keep more assets on "standby" than other asset-intensive industries do. Equipment such as x-ray machines, MRIs, and operating tables needs to be available on an as-needed basis for people with acute and emergency needs. For example, prior to the COVID-19 crisis, many hospitals had four to five times the number of ventilators than would be needed on a typical day—yet there was a massive, nationwide shortage of ventilators during the initial months of the crisis. Hence, having a buffer of spare assets is a perfectly reasonable mandate for healthcare organizations, regardless of the lower asset utilization that it entails.

HIGHLY SPECIALIZED ASSETS

A third factor is that many expensive assets in healthcare (e.g., linear accelerators, proton beam machines) are highly specialized and perform only a very narrow set of services. There aren't enough local patients in need of such services to keep these machines busy all day, every day. A full-service health system needs to invest in some of these assets, knowing full well they will remain idle for several hours each day.

The three above factors are somewhat baked into the current system and may be difficult, if not impossible, to change. However, there is a fourth factor that is contributing to poor asset utilization in medicine:

OUTMODED SCHEDULING PROCESSES

The processes that providers use for scheduling patients, procedures, and equipment have remained largely unchanged for decades, despite the exponential increase in the complexity of healthcare services.

Unlike the first three factors, this is a problem that *can* be addressed. There *is* a better way of handling appointments in healthcare, one that can help maximize asset utilization *and* minimize patient waiting times. But it will require embracing some new paradigms and overcoming the mindset of "This is the way we have always done it."

Before we talk about that new approach, we would like to briefly share the story of how we, the authors, "chanced" upon the idea of transforming healthcare operations in this way and achieving *Better*

Healthcare through Math.

THE LEANTAAS STORY

LeanTaaS, the company for which we both serve as senior executives, was founded in 2010 with the idea of combining Lean principles for operational excellence with advanced data-science algorithms and delivering them as a "software-as-a-service" suite of products. Software as a service means that, instead of buying a software package outright and installing it on their local hardware system, our customers pay a monthly fee to use our cloud-based products, which are accessible anywhere on any device.

Initially, we didn't focus specifically on healthcare—we built and refined our technology platform and business model by helping customers in a wide variety of industries, such as banking, insurance, retail, technology, manufacturing, hospitality, and consumer goods. Our specialty was solving operationally complex problems, no matter what the industry was. One of our projects, for example, was optimizing the tool-rental system for The Home Depot.

One particularly tough challenge we faced early on was helping a casino manage its complex schedule of dealers and tables—a situation where putting a rookie dealer at the high-roller table could cost the casino a million dollars in losses on a weekend night. This casino operated its gaming tables twenty-four hours a day. It offered about twenty different games, each with its own scheduling demands and multiple tables. So, for example, at 3:00 p.m. there might be three blackjack tables open, at 6:00 p.m. three more tables would open, and at 9:00 p.m. all blackjack tables would move into full swing. A dealer needed to be assigned to each table, and that dealer needed to know how to play the game at that particular table. Yet when it came to employee scheduling at this particular casino, the managers were just winging it and assigning dealers based on "I don't want to work Friday night this week" criteria.

This project taught us a lot about optimizing operations by balancing the demand-and-supply signals in a complex, high-stakes environment. As a result of this and other challenging projects, we were able to build some powerful algorithms that allowed frontline workers to employ our math and make better decisions *automatically*.

In late 2013, Sridhar Seshadri, DBA, the chief administrative officer for destination service lines at Stanford Health Care, approached LeanTaaS to help solve a critical operating challenge. The waiting room in the hospital's infusion treatment area was regularly overcrowded by midday, but nearly empty during the early morning and late afternoon. Stanford Health Care needed help redesigning its operating model to be more efficient while still providing the excellent care and service for which it is known.

This was an exciting challenge for us—a chance to combine sophisticated data-science algorithms with proven Lean principles such as Heijunka—or "level loading"—to reduce waiting times for infusion services for Stanford Health Care's Cancer Center.

What is level loading? For a simple illustration, think about freeway traffic around a major city. If you head out on the highway in the early morning, you'll notice there's hardly any traffic on the road. From around 6:00 a.m. to 7:00 a.m., traffic starts to ramp up quickly, and then, between about 7:00 a.m. and 9:00 a.m., there's a predictable period of bumper-to-bumper gridlock every day.

Now let's pretend for a moment you could get 10 percent of the commuters to work a 7:00 a.m. to 3:00 p.m. shift, 10 percent to work from 8:00 a.m. to 4:00 p.m., 10 percent to work from 9:00 a.m. to 5:00 p.m., and so on. So you would distribute 100 percent of the people who work in the city into ten equal buckets. By doing so, you would achieve level loading on the freeways, meaning there'd be approximately the same number of cars on the lanes every hour of the day. The freeway would thus be optimally employed, and the drive time for everyone would be reduced by more than half. Sadly, you can't do that with highways. But you can with healthcare.

Over a period of nine months, the LeanTaaS team worked hand in hand with Dr. Seshadri's team and developed an optimized template for each day of the week that provided specific guidelines on the number of slots that could be offered for each type of appointment at every ten-minute interval throughout the day.

By the fall of 2014, it became clear that our approach was working. Patients were not being forced to wait nearly as long for their infusion treatments. One patient even commented that she had stopped bringing a book along with her because she was now confident her treatment would begin at the scheduled time.

We began to wonder whether the scheduling problem we solved at Stanford Health Care was unique to Stanford or whether other infusion centers around the country might be experiencing similar issues.

Not having any real market research capacity, we did what any scrappy start-up would do: begged our friends for help. We called two former colleagues—Felipe Osorno, who led the value improvement office team at Keck Medicine of USC, and Scott Lichtenberger, who was the chief strategy officer at UCHealth in Colorado. Both were gracious enough to facilitate meetings with the senior executive teams of their respective institutions.

Within a few months, we were batting three for three; both USC and UCHealth agreed to move forward and pilot our approach in their infusion centers. In both of these additional settings, we were able to demonstrate a similar level of success—a 10 to 15 percent increase in the effective capacity of the infusion area combined with a 30 to 40 percent reduction in the wait times for patients during the midday "rush hour" of 11:00 a.m. to 2:00 p.m. We found a solution to a complicated problem that could be scaled—initially to infusion centers, but also, we were convinced, to many other assets within the health system. We knew we were on to something.

Over the next eighteen months or so, we systematically transitioned all of our nonhealthcare software applications to third-party partners so that LeanTaaS could focus 100 percent of its energies on solving what we believe to be a hugely important problem—matching the incoming demand for healthcare services with the available capacity of the relevant assets (people, equipment, rooms) throughout the day, every day of the week.

We have been pursuing this vision with a single-minded focus since 2015. This narrowing of our focus has been the best thing to happen to us as a company. As of December 2019, it has enabled us to deploy our products at more than a hundred health systems, raise over $100 million in funding from health systems and venture capital firms, grow our team to over two hundred employees in the United States, and sharpen our thinking on the operational principles, the underlying mathematics, the data science, and the optimization and simulation algorithms, all of which are vital to consistently delivering real and measurable improvements to the operational performance of health systems.

So now let's take a deeper dive into the principles we use to help healthcare systems become consistently excellent in their operational performance. As promised, we'll steer clear of the mathematical "black boxes" of predictive analytics, machine learning, and algorithms, but it is important that you understand some of the concepts *behind* the math.

Please, have a seat. This won't hurt a bit.

CHAPTER 2

BETTER HEALTHCARE THROUGH MATH

We hope you agree with our premise thus far: healthcare is facing mounting financial pressures that render "the old way of doing things" no longer ideal as a long-term operational solution. One of the crucial steps the industry must take is to optimize its asset utilization—despite some of the inherent limitations we discussed, such as the unavailability of many of these assets for two-thirds of the twenty-four-hour clock. (And when we talk about assets, remember, we are not referring only to expensive pieces of medical equipment, but to human assets as well.)

In some industries, asset utilization is a fairly straightforward thing. In traditional manufacturing, for example, we can accurately determine the number of widgets to be made in a given day. That is because we understand exactly how long it takes the machine to stamp out each widget, and we have a high degree of control over the variables, such as the personnel, supplies, and maintenance steps needed to keep that machine humming all day. Hence, we can push for 90 or 95 percent machine utilization on an average day and use the planned downtime to perform preventative maintenance on the machine to ensure its high uptime reliability.

Unfortunately, healthcare doesn't work that way. The number of patient appointments on a given day is highly variable, and the exact services each patient will need, as well as the time it will take to deliver each of those services, is difficult to predict. Finally, the availability of the appropriate staff member, room, supplies, and equipment is heavily dependent on the aforementioned parameters and is highly variable as well.

TWO FOUNDATIONAL CONCEPTS: MATCHING AND LINKING

Improving asset utilization in healthcare, therefore, requires a deep understanding and management of two foundational concepts:

1. **Matching**: A fixed supply of resources—each constrained by real-world challenges—must be matched with an unpredictable, volatile demand for those resources. A moving archer must hit a moving target, often within a narrow time window.

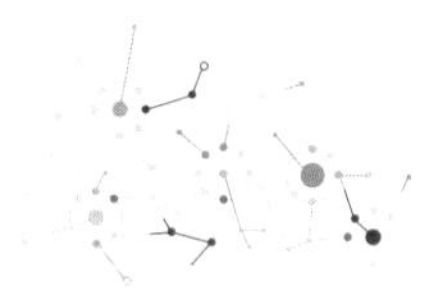

2. **Linking**: Individual healthcare services (e.g., lab visits, consultations, medical treatments)—each of which has its

own inherent variability issues—must be strung together to provide an end-to-end patient experience that is as seamless as possible.

To harness these two concepts in an intelligent manner requires more than hope, luck, and a calendar. Let's look at each of these concepts in turn.

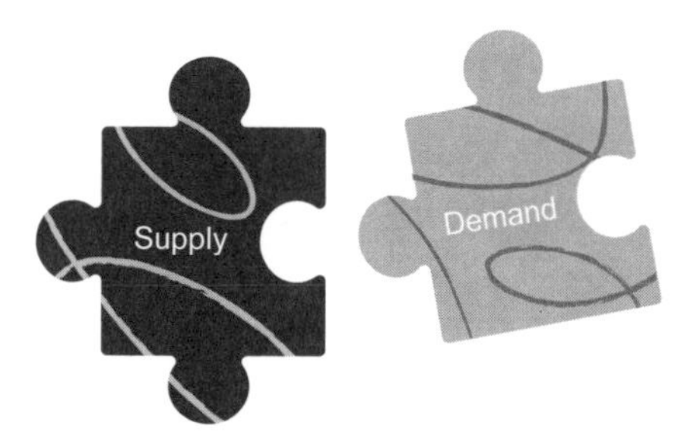

MATCHING DEMAND AND SUPPLY

Demand for any medical service is difficult to precisely predict. The "demand signal" includes multiple factors such as the number of patients who will arrive at the facility on a given day, the type of treatment they will need, the time of their arrival, and the probability of a late arrival—or, in the case of operating-room surgeries, the number of elective and add-on cases that have been scheduled that day.

This demand signal, as we have pointed out, is highly variable and diverse. The volume of patients varies substantially based on the day of week and the time of day, and there is often a complex mix of short and long, simple and complicated appointments. When a provider sits down with a patient, he/she has no way of knowing whether this patient will need additional services or how involved those additional services may be. Cancellations, add-ons, delays, and no-shows are additional factors. Outside events, such as major accidents, natural disasters, outbreaks of illness, and weather, can also affect the demand signal for many medical providers.

Supply, as we have seen, is limited, constrained, and also unpredictable. The resources needed to execute any one particular medical appointment can include: the right staff member with the right skills to conduct the procedure; the right facilities, such as a room, chair, or bed; the right equipment, such as a surgical robot or infusion pump; the right supplies, such as the correct mix of drugs for a particular chemo patient; and the right ancillary services, such as labs, imaging, pharmacies, physiotherapy, etc.

Supply is additionally constrained by factors such as the availability of the right room for a specific surgical case, the schedule/attendance of staff, the current workload of the staff, the specific skills required for each aspect of the appointment, and the utilization of shared assets such as procedure rooms or imaging machines. These shared assets are used by other providers in the area, much in the same way that runways at an airport are shared by all of the airlines or highway lanes are shared by all cars. Individual providers often have little or no control over the hour-to-hour availability of shared resources. In the case of operating rooms, capacity is typically reserved as "block time," which is a bit like reserving airport runways by airline, hoping they will be used efficiently.

The matching of supply with demand needs to happen within tight time windows every single day so that the right patient is steered to the right resource at the right time.

The matching of supply with demand needs to happen within tight time windows every single day so that the right patient is steered to the right resource at the right time. This is a tricky challenge that requires sophisticated solutions.

Unfortunately, the way most medical appointments are made

today is as follows: two people—the patient and the scheduler for the clinic (or, in the case of surgeries, the clinic scheduler and the OR scheduler)—look at calendars and agree on a mutually convenient slot. End of story.

But there are better methodologies available. These improved methods enable providers to more accurately place the right patient in the right slot at the right time, thus helping to maximize patient access, reduce patient wait times, increase asset utilization, better utilize the OR, and reduce costs. These new methodologies, alas, are not the kind that can be "manually operated" or employed in a "seat-of-the-pants" manner. They involve complex math algorithms and AI. However, this powerful math *can* be harnessed in the form of easy-to-use online tools that frontline staff can comfortably employ. These tools not only help to optimize asset utilization but also make the jobs of frontline scheduling staff dramatically easier. They can make finding an optimal appointment slot as easy as booking a dinner reservation on OpenTable.

It is important to make a clear distinction between *scheduling* and *optimization.* Scheduling is the act of putting down a specific appointment onto a calendar, regardless of whether the calendar is on paper, on a spreadsheet, or on an online calendar of some sort. Optimization is the underlying intelligence that determines the best option for scheduling a particular type of appointment or allocating a specific type of asset based on the sophisticated consideration of dozens of factors that influence both the supply side as well as the demand side of that decision.

Technology that embeds optimization into online scheduling tools is ubiquitous in today's sophisticated business environment, where profitability depends upon executing a high volume of transactions while maintaining a high level of asset utilization.

HOW UPS DOES IT

Let's consider the complexity of UPS's demand-supply matching problem:

On the demand side, UPS delivers nearly sixteen million packages a day[16] without ever being able to know, in advance, precisely how many packages will need to go from any specific location to any other specific location, worldwide. Many of these packages are overnight shipments from far-flung cities and need to be delivered to specific addresses before 10:30 a.m. the next morning.

Supply-wise, UPS Airlines owns, leases, or charters over 550 aircraft[17] and is by far the largest airline in the world, based on the number of destinations served. The aircraft are positioned in hub cities, where a vast network of UPS trucks collects packages from thousands of route vans, each of which stops at 120 locations per day, on average,[18] to deliver inbound packages and gather outbound packages before the close of business at 5:00 p.m. local time.

These packages are then processed in sorting hubs and sent on to long-haul trucks for ground delivery or loaded onto aircraft for long-haul air delivery. The entire process works in reverse after the planes land and the long-haul trucks arrive. The incoming packages are sent to sorting hubs, where they have a very tight "sort span" window (three to four hours between midnight and 4:00 a.m.) before being loaded onto trucks again and making their way to local hubs, where they are stacked in an organized manner within the familiar brown parcel vans we see cruising our streets.

---

16 "UPS Fact Sheet," accessed July 2020, https://pressroom.ups.com/pressroom/ContentDetailsViewer.page?ConceptType=FactSheets&id=1426321563187-193.

17 Ibid.

18 Marcus Wohlsen, "The Astronomical Math Behind UPS' New Tool to Deliver Packages Faster," Wired.com, June 13, 2013, https://www.wired.com/2013/06/ups-astronomical-math/.

The sorting hubs are a sight to behold—hundreds of dock doors where long-haul semis unload their cargo onto a spiderweb of *dozens of miles* of conveyor belts that scan the packages and automatically kick them onto the belt destined to load a specific truck waiting at a specific dock door.

All of this would be impossible, of course, without sophisticated demand-supply matching to ensure that the right number of short- and long-haul trucks and aircraft are positioned in the right place at the right time every day of the year, including vacations and holidays—when the patterns change. Elaborate route-planning methodology is needed at each local hub to ensure that the fifty to sixty local parcel vans can cover the entire delivery area in a smooth and efficient manner. UPS drivers have tightly defined rules about following speed limits, never backing up, making three right turns instead of a left turn to minimize idle time, and handling packages in a way that minimizes the time spent at each stop. If every UPS driver were to add even one unnecessary mile to their daily route, that would cost the company $30 million a year.[19] Thus, efficiency has become a true science at UPS.

HOW UBER DOES IT

Uber is another company that faces intense demand-signal challenges. The demand for rides is highly variable, based on the time of day, day of week, month of year, and zip code. The demand is also quite *diverse*—some people require UberX service, while others need Select, XL, Black Car, SUV, wheelchair accessible, Uber EATS, and more. Additionally, demand is unpredictable because there are always sporting events, weather events, conventions, public assemblies, and the like that make it hard to predict the demand precisely.

---

19 Ibid.

On the supply side, Uber faces many problems that are even more uncontrollable than those in healthcare. Drivers aren't salaried employees. Each day they can decide independently whether they will work or stay home; Uber has no leverage to *force* drivers onto the road. Also, at any given time of day or night, there may be the wrong number of small cars versus large cars in a certain area. Or there may be traffic snarls in some spots, making it difficult for drivers to reach riders. Too many drivers may be operating in some geographical areas, and not enough in others. So Uber's supply variables are extremely volatile.

Consequently, Uber devotes a tremendous number of resources to predicting both the demand and the supply for every geographic market every *minute* of every day while maintaining a real-time focus on any imbalance between the two.

When the system detects a shortage on the supply side, it might ping more drivers to come out and drive. When it detects *too many* drivers in a particular area, it might reposition some of the drivers a few miles in any direction to spread out the supply more favorably.

On the demand side, it may offer riders credits during light-demand hours to stimulate more customer usage or enforce surge pricing during busy hours to decrease the demand signal and bring it in line with the available supply.

Matching the supply-demand signal in this manner enables Uber to consistently deliver a wait-time experience that is usually well under ten minutes in most metro areas at most hours of the day or night. Achieving such short wait times for millions of daily customers is an astonishing logistical feat when you consider all of the variables in play.

The "technology stack" behind this feat is extremely sophisticated and built by teams of product managers, mathematicians, scientists, and software engineers over many years. It analyzes and predicts ETAs, drive times, optimal trip routes, rider patterns, fees, traffic issues, and more. The system must continually *learn* as well, so that every miscalculated ETA and missed driver-rider connection helps Uber refine its calculations even further.

Imagine if Uber tried to do all of this "manually"—that is, to run its service based on looking at open slots in a schedule and trying to pair riders with drivers at mutually convenient times and locations. That is essentially what traditional taxicab services did, and that is why Uber and Lyft have been gaining market share with such dizzying speed.

We believe healthcare institutions that make the effort to deploy these advanced methods will see remarkable improvements in their operational performance (e.g., shorter calendar waits for appointments and shorter stays in waiting rooms), which will lead to an improved patient experience and better competitiveness in the marketplace.

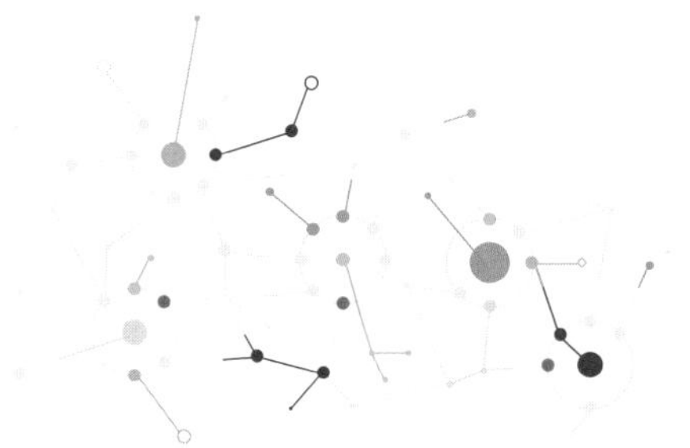

LINKING INDIVIDUAL SERVICES

The other critical concept underlying asset utilization at healthcare companies is the linking of individual services together. Many patients undergo multiple services as part of each encounter with a health system. For example, they often visit the lab to get their blood analyzed, then consult with their provider, and then proceed to receive an additional service (e.g., an infusion treatment, physical therapy, or a calibration of a personal medical device).

Scheduling linked appointments is particularly challenging because:

- The individual appointments must be spaced closely enough to be convenient for the patient but far enough apart to leave a sufficient buffer of time to accommodate the delays that are likely to occur.
- Each of the linked services is typically scheduled by a different department, and there isn't good visibility into the availability of the assets within the other departments.
- Any of the appointments, either with the main provider or with any of the linked departments/providers, may run long due to complications with the procedure and/or the patient.
- Billing/payment/insurance issues might arise.
- There may be patient overloads in any of the individual departments, which then have a domino effect on all the other linked appointments for all of those patients.
- There are often logistical considerations that add to the complexity of scheduling linked appointments. Examples include the walking (or driving) distance between the various services involved in a single patient encounter and the need for special services such as transportation or language translation.

HOW AIRLINES DO IT

The airlines are masterful at solving the dual problems of matching supply with demand *and* linking services together efficiently. Although air travel was challenged and changed by COVID-19, it does mirror healthcare in some significant ways. Consider the sequence of events that takes place while you are waiting to board your flight:

The inbound plane arrives at the gate. The ground crew staff, armed with their orange glow sticks, guide the incoming aircraft into position using marks on the tarmac. As soon as the plane stops at its gate, the jet bridge is moved into place and connected to the aircraft, and the incoming passengers start to make their way into the terminal.

Over the next forty-five minutes, a flurry of parallel activities takes place. Some of these activities are "above-wing" services such as deboarding, cabin cleaning, restocking of food and beverages, and boarding, while others are "below-wing" services such as refueling, baggage handling, and checking engines.

It is helpful to think of each of these activities as its own "service node." Each service node is an independently functioning unit that has been mathematically optimized using the demand-supply balancing concepts we've been discussing. As examples, consider the following:

- The right number of baggage carts and baggage-handling personnel needs to be on hand to unload the correct bags that were on the incoming flight and then load the bags for the outbound flight within a fifteen-to-twenty-minute period.
- The exact right number of cabin-cleaning personnel must be deployed to quickly and efficiently clean the cabin and the restrooms on this particular aircraft prior to its departure.
- The supply trucks with the snacks, drinks, and meals must be in the right place at the right time and correctly loaded.
- The correct number of maintenance technicians for this particular type of aircraft must be standing by to quickly perform the needed safety and equipment checks.

- A fuel truck with the correct amount of fuel must refuel the aircraft efficiently.
- A correctly sized tug vehicle for pushing the departing aircraft back from the gate for departure must be positioned nearby and ready to move into place.

The calculation that goes into determining the number of each asset type (fuel trucks, baggage carts, tugs, etc.) and the staging location for each is based on a detailed understanding of "pushes per fifteen minutes." This is a measure of the number of aircraft pushbacks (i.e., the pushing of a plane, via a tug, backward from the gate) that occurs within each subsection of the airport.

Why is this type of calculation needed? Because some of the needed ground support vehicles don't travel at high speeds, and major airports span several miles from end to end. Therefore, it is imperative to have the right number of support units in the right proximity of their job area at every point in time throughout the entire sixteen to eighteen daily hours for which most major airports operate. This is similar to Uber's need to have the right number of drivers circulating in a given area to accommodate the expected rider demand in that area.

Now consider the linkage issue. All of these various services—the cleaning, the refueling, etc.—must be precisely choreographed with one another in order to execute a complete turnaround of a large aircraft within forty-five minutes. This requires an appreciation of several factors such as:

- The cabin cannot be cleaned until the passengers have disembarked, but it *must* be cleaned before the new passengers come on board.
- The incoming bags (destination bags and transfer bags) must

be unloaded from the cargo hold before the outbound bags can be loaded.

- The baggage carts, fuel trucks, supply vehicles, and deicing truck (when needed) must navigate the cramped area under and around the wings without colliding with one another or with the aircraft.
- The supply truck for food, snacks, and drinks must dock at the opposite door from the one used by exiting passengers so that it can get started on restocking the aircraft without having to wait for all the passengers to disembark.

Matching (supply with demand) and *linking* (connected services) are core aspects of daily operations for the airlines. Delta Airlines alone has over a thousand weekday departures from its Atlanta hub. The only way it could possibly orchestrate this sheer quantity of service is by tailoring each service "node" to be rightsized and operationally efficient, and by coordinating all of the individual services with one another so they can be completed within the tight forty-five-minute turnaround window. To accomplish this requires a mastery of matching and linkage.

And for those competitors that learn to be *hyper*efficient with both of these service aspects, additional profits await. Southwest, for example, is particularly fast at turnarounds and usually accomplishes them within twenty-five to thirty minutes. This added level of efficiency gives the airline a utilization advantage of one additional flight leg per day, on average, for each plane in its fleet. Southwest currently has 745 aircraft in its fleet (all 737s),[20] so an extra leg per day for each plane adds up to more than a quarter million potential additional

20 Southwest Airlines 2020 Corporate Fact Sheet, https://www.swamedia.com/pages/corporate-fact-sheet.

legs per year! Southwest has a fixed number of assets and a fixed number of hours in a day, just as all the airlines do, but its superior asset utilization paves the way for increased revenue and profitability.

And better turnaround time is only *one* way airlines have increased asset utilization. They have optimized their flight schedules and ticketing systems as well. The airlines have become extremely sophisticated with the mathematics of filling seats—known as yield management, a system that segments future seats on a particular flight into price bands that dynamically grow and shrink based on the actual bookings as compared to the projected ones. At the same time, booking seats is easier, more transparent, and more affordable than ever for passengers. This is the kind of win-win scenario toward which healthcare ought to aspire.

APPLYING THESE CONCEPTS TO HEALTHCARE

As UPS, FedEx, Uber, Lyft, and the airlines have been doing, healthcare companies must understand the importance of optimizing both the matching and the linking of the various components of service delivery.

To picture this properly, we can think of each service (labs, clinic visit, procedure) as a node in a network—as we did with the airline example—and the flow from one service to the other as an edge between two nodes. The most effective approach to optimizing such a network is to optimize the nodes *before* attempting to optimize the edges. FedEx, UPS, and Amazon streamline their warehouses before they worry about making their drivers drive faster or using drones to speed up deliveries.

The first step in optimizing any asset is to develop a realistic estimate of the target utilization for that asset. When a system approaches its maximum utilization, it goes into gridlock. To understand this, we come back to the concept of rush-hour traffic on the freeway. During rush hour, the freeway is almost fully utilized—it is as packed with cars as it can possibly be. As a result, all of the key metrics suffer to a stunning degree. A drive that normally takes fifteen minutes takes ninety minutes at rush hour, and an accident that could have been cleared in ten minutes now takes two hours to clear.

Therefore, the trick to optimizing each asset is to prevent it from going into gridlock in the first place—which implies that it needs to operate at a level that is appropriately *below* maximum utilization for as many hours of the day as possible. The more variable the demand signal, the more buffer you must leave between your target utilization and 100 percent utilization.

So although the finance and administrative leaders of a health system might want to strive for 90 percent utilization of their most expensive assets, such a utilization rate is unrealistic in healthcare and likely to cause major operational headaches. The booking of healthcare assets must retain some flexibility to be responsive to the emerging needs of patients as they arise in real time throughout the day.

The trade-off for this needed flexibility is lower utilization rates. This is an easy concept to grasp using a simple analogy. Most of us want the flexibility to be able to jump into our cars anytime, day or night. In exchange for this convenience, we accept the fact that the utilization of our cars is probably going to be less than two hours per day on average (i.e., less than 10 percent utilization). If we wanted our cars to be more fully utilized, we would have to agree to share them with other families on our street and then carefully work out the logistics so that everyone could get where they need to go, on time, each day. But the greater the variability and unpredictability of the destinations (distances) and departure times of the other users, the greater the likelihood that people will end up waiting for long periods of time for a car to return and become available. Therefore, most of us have accepted the reality that our cars will indeed have a low level of asset utilization—but that we are OK with it.

The bottom line is this: we want our medical assets to be as fully utilized as possible *without* going into gridlock and/or causing long waits for patients—knowing as we do that such delays also have a knock-on effect on the patients' linked appointments. The ability to strike the ideal balance between underutilizing and overutilizing assets, across an entire system, requires not only a willingness to tackle this logistical issue with eyes wide open, but also a level of math calculations that simply cannot be performed manually—even with the best instincts or judgment.

The ability to strike the ideal balance between underutilizing and overutilizing assets, across an entire system, requires not only a willingness to tackle this logistical issue with eyes wide open, but also a level of math calculations that simply cannot be performed manually—even with the best instincts or judgment.

When you add in the complication of linking appointments together efficiently, the math becomes even more unwieldy. Under the present system, schedulers make only the most primitive, shoot-from-the-hip attempts to provide good linkage. When scheduling a patient for an infusion visit, for example, the scheduler hopes that by giving the patient a 7:30 a.m. lab visit, an 8:00 a.m. provider appointment, and a 9:00 a.m. chemotherapy session, things will simply "work out," or the downstream staff will do whatever they can to "make it happen."

Adding stress to this system is the fact that providers often exert influence on the downstream units. It is not uncommon for a provider to say, "I have a patient scheduled to see me at 8:30 a.m.;

I'd like her to start her infusion treatment at 9:15 a.m." This type of unilateral demand makes no sense whatsoever in light of the overall efficiency of the system. What if twenty other providers did the same thing? The infusion unit would be stymied and unable to offer any appointments at all, knowing that providers might give away all their slots without even consulting them.

Imagine the pilot of an incoming aircraft telling ground operations that he has twenty passengers on board who need to make a connecting flight to New York and would therefore like to have the departure gate for the New York flight reassigned so that it is adjacent to the gate where he's landing. This kind of decision-making power in the hands of individual pilots, who are not able to see the big picture, would throw the whole system into chaos.

The humbling truth is that we human beings individually lack the mental computing power, the holistic perspective, and the data necessary to make the kinds of decisions that lead to optimal *system* functionality in today's technologically complex world. We require help. To even *approach* the level of sophistication at which the ride-sharing, package delivery, and airline companies operate, health systems must employ modern tools. And these tools must be far more robust than the existing suite of dashboards, homegrown analytics, Excel spreadsheets, calendars, and manually generated reports.

"But isn't that why we all worked so hard to adopt EHRs?" we hear you asking. Yes, it's true that enormous investments have been made in the deployment of electronic health records, and these tools have been hugely beneficial, but EHRs simply are not designed to do the kind of predictive and prescriptive analytics we're talking about here. It turns out the EHR is a necessary but insufficient tool to address the complexity of managing healthcare assets efficiently.

Let's look at why ...

CHAPTER 3

YOUR EHR AND TRADITIONAL TOOLS AREN'T ENOUGH

Over the past fifteen or twenty years, the major EHR software vendors have made an enormous contribution to healthcare and society by digitizing patient information into electronic health records. In addition, almost every health system has invested in dashboards, either internally built or as add-ons to packaged software solutions, to help their staff make operational decisions.

To paraphrase many of the hospital leaders we work with:

> *We've spent tens (or hundreds) of millions of dollars on EHR implementation, and on top of that, we have invested heavily in reporting capabilities. We have dashboards throughout the hospital to keep track of everything, and teams of people dedicated to BI, reporting, data visualization, ETL, and custom report*

generation. We have even installed command centers. Yet we haven't really moved the needle when it comes to improving operational performance.

Unfortunately, EHRs and the standard reporting tools are not enough. Relying on them alone can often result in "admiring the problem" instead of solving it. Worse, these methods can have you looking at the wrong problem. Why? Because the metrics being used often don't make sense (e.g., block utilization, turns per chair) or are not believed internally—we have seen many Excel spreadsheets floating around hospitals with definitions of "turnover time" and "block utilization" that disagree with one another.

Let's look at the EHR first.

WHY THE EHR ISN'T ENOUGH

For the sake of clarification: sometimes the terms *EHR* and *EMR* are used interchangeably, but really they are two different things. The EMR (electronic medical record) came along first. It was essentially a digitized version of the patient's paper charts. An EMR can help clinicians diagnose and treat patients and also track their treatments and symptoms over time. But it isn't really designed for sharing with other providers, and it isn't particularly portable.

The EHR (electronic health record) is based on a grander vision. It is envisioned to be the future of healthcare documentation and communication—a universal record that contains not only comprehensive clinical data for the patient, but also other vital information such as demographic data, insurance information, input from patient medical devices, and access to tools providers can use for better treatment and decision-making. The EHR aims, ideally, to represent the *total health picture* for a patient. Perhaps most significantly, it is designed to facilitate the sharing of real-time input from all of the patient's providers and to "travel with" the patient from practice to practice. Portability and universal access are its essence.

In those health systems that have implemented EHRs, the EHR has quickly become the single source of truth. It contains all the relevant clinical information about every encounter the patient has had with the system and serves as the one and only repository for all test results, medication histories, provider notes, diagnoses, allergy information, lab data, and records of clinic visits and procedures, among other things.

EHRs have also become integral to the scheduling process. By showing users all the available resources in the health system, they enable patients to be scheduled simply by treating each appointment as a "reservation" of a specific resource at a specified time. They also allow providers to gather data about past appointments and pull statistics about "global" things like the prevalence of specific medical complaints at specific times of the year.

The EHR arrived on the scene with the promise of a new digital frontier. Health systems have spent many years and hundreds of millions of dollars implementing EHRs and training their staff in using them. (An average multiprovider clinic will spend about $162,000 in EHR implementation costs,[21] and a large academic medical center could spend as much as $250 million deploying an EHR.) So healthcare operators, justifiably, have a difficult time understanding why the EHR cannot do more to help them optimize the utilization of their assets or improve the flow of patients across

21 Jeff Green, "How Much EHR Costs and How to Set Your Budget," EHRinPractice.com, updated March 7, 2019, https://www.ehrinpractice.com/ehr-cost-and-budget-guide.html.

the various interconnected services that make up a single, complex encounter between the patient and the health system on any given day.

> The reality is that EHRs simply do not have the mathematical underpinning to help solve complex optimization and utilization problems.

The reality is that EHRs simply do not have the mathematical underpinning to help solve complex optimization and utilization problems. Here are four specific ways in which the EHR is inadequate for solving the types of problems we've been discussing so far.

1. EHRS ARE BUILT ON A FLAWED SCHEDULING ASSUMPTION

EHRs are built on a universal and simplistic assumption that has long been baked into the scheduling mindset of the healthcare establishment—that is, that scheduling an appointment is equivalent to "reserving a resource." Thus, if John Doe has an MRI appointment from 8:00 to 9:00 a.m. on a given day, then a specific physical (or virtual) machine/resource—let's say MRI machine 1—must be reserved for that time slot. And so the staff member dutifully enters John's name or medical record number (MRN) into that space on the schedule, and that block now belongs to John Doe.

This explains why grid-based schedules are so omnipresent in health systems. These grids may appear in the form of hard-copy printouts, calendars, Excel spreadsheets, handwritten whiteboards, digital Snapboards, and/or TV displays, but they all function in essentially the same way. They list the specific assets (e.g., chair, room, machine, or provider) across the top of the grid, and the times of day down the left side of the grid. Whenever an appointment is confirmed, the corresponding block on the grid is colored in. Simple, visual, intuitive.

Also completely inadequate.

The block-grid is an excellent approach to building a daily schedule for the assignment of tennis courts or conference rooms. That's because tennis sessions and meetings run on a time interval that is *deterministic*—the session being reserved has a known start time and a known end time *at the time the appointment is made.* A deviation in either the start time or the end time is not really a problem: the individuals using the resource simply get kicked off the tennis court or out of the conference room when their allotted time is up. Throughout the day, the holders of each scheduled block surrender the resource at predictable, timed intervals.

By contrast, clinical appointments are *stochastic*—inherently random and unpredictable. The start time and end time of a treatment or consultation session can never be precisely known at the time of making the appointment. We all pretend that it can for scheduling purposes, but in practice it cannot. If the 8:00 a.m. patient's appointment runs long, the *8:30 a.m.* patient's appointment starts late, period. And there is no recourse for *forcing* things back on track once the day has started—you cannot kick a patient out of their treatment midstream just because their treatment runs past its scheduled end time.

The only time a grid schedule, with all of its beautifully colored blocks, attains perfection is in the precious moments before the practice opens in the morning. Once the day gets rolling … life gets in the way. Appointments are canceled or delayed, patients are added onto the schedule by providers, the staff for one or more of the essential services—such as the lab or pharmacy—gets backed up. The frontline staff has no choice but to engage in a seat-of-the-pants game of Tetris, trying to fit the new and emerging puzzle pieces into place, shuffle the remaining appointments as efficiently as possible, and somehow reduce the chaos in order to survive the day.

> **The real problem is that we are using a simplistic and outdated model to address the complex issue of matching patients to resources.**

And just as in the movie *Groundhog Day*, the grid for tomorrow will produce similar results.

We tell ourselves this flawed process "works," but only because we have come to accept that long waiting times, stressed staff, frustrated providers, and a certain level of office chaos are inevitable aspects of running a clinic. But the real problem is that we are using a simplistic and outdated model to address the complex issue of matching patients to resources. EHRs, alas, are built on this old model.

2. EHRS EMPLOY THE "HEY, HERE'S AN OPEN SLOT" PRINCIPLE

EHRs encourage health systems to fill their schedules on a first-come-first-served basis. Hence, schedulers fill up their calendars with appointments in the order that the requests come pouring into their inboxes or phone queues. If there's an open slot, and the patient wants it, the patient receives it.

Again, the tools and methods most health systems use to set up appointments simply ignore the inherent complexity as if it did not exist: the scheduler and the patient find a mutually agreeable slot and then place the appointment in that slot. Period.

From an optimization and asset-utilization point of view, this makes absolutely no sense. It is akin to letting individual pilots prebook their runway access at a busy airport weeks ahead of time, using a calendar and a pencil. "No one is scheduled to land on Runway

34-R between 1:00 and 1:10 p.m. on Thursday the eleventh, so I will reserve that time now." That would be absurd, given the realities of air traffic—the delays, the cancellations, the wind and weather. A busy healthcare clinic is not so different.

Here is another analogy. Imagine assembling a jigsaw puzzle like this: two people pick up a jigsaw puzzle piece and look for a mutually agreeable spot on the table to place it. They don't know what picture they are trying to form, and they don't look at the other pieces; they just choose an uncluttered spot and place the piece there. Now two more people come along and do the same thing with the next puzzle piece, and so on. There is no chance that the final outcome would be a correctly solved puzzle with all the pieces fitting together snugly. Similarly, there is no chance that the appointment book created for any given day in a busy medical office by using an individualized first-come-first-served approach is likely to be anywhere near the optimal solution.

When it comes to OR (operating-room) scheduling, many community clinics are unable to look at the OR schedule directly because they don't have access to the EHR. So when clinics need to book time outside of their assigned block, which happens frequently (as we will see in Chapter 4), what often follows is an endless back and forth between the clinic and OR schedulers to fit a patient in. Even if the clinic *does* have access to the EHR, and they see an open slot on the schedule, they still have no way of knowing if that slot actually works, because the right combination of room, equipment, surgical case team, and anesthesia may not be available.

Healthcare appointments must be *intelligently sequenced* (with the right logic built in, including the availability of the right resources) in order to keep the asset as busy as possible—without gridlock. And in order to perform such intelligent sequencing, there must be a mathematical understanding of how all the dynamically moving parts fit together in optimal ways.

3. EHRS DON'T HAVE THE NEEDED MATH ENGINE

EHRs do not incorporate any of the sophisticated concepts used by scheduling-intensive businesses (Amazon, Lyft, the airlines, etc.) to optimize their assets—such as probability, simulation, and machine learning.

The airlines, for example, use vast amounts of historical and statistical data to estimate the number of passengers that are likely to show up on time for their flight, and to determine the probabilities for full flights. Knowing there is only a remote statistical chance that all two hundred booked passengers will show up for a two-hundred-seat flight, they routinely overbook their flights. They understand there will be cancellations, add-ons, no-shows, etc. By overbooking, they are able to maximize the utilization for each aircraft on each leg, rather than allowing a mathematically predictable number of seats to go empty.

Many people object to the overbooking system as a bad business practice, but for the most part, the system works pretty well for everyone, and standby passengers succeed in getting onto their flights. On the rare occasion the airline gets its calculations wrong, it offers incentives at the gate to encourage a few volunteers to give up their seats. Typically this is not a problem, and several passengers are willing to take a later flight in exchange for the perks they receive. If the airline had not overbooked, it would have flown with some empty seats (i.e., underutilized assets), which would likely result in fare increases. So, in a sense, "everyone wins" under the current system—at least the vast majority of the time.

Health systems would be wise to employ such sophisticated approaches—after all, even if they err a bit, patients will only have to wait a few extra minutes (and probably wouldn't even notice, as accustomed as they are to waiting for their appointment).

However, the EHR simply does not allow the scheduler to schedule three appointments into two slots or to schedule a four-hour treatment into a three-hour slot, even when the EHR's own data suggests a high probability that not everyone will show up as scheduled. Hence, the EHR forces the frontline staff into "betting against the house" with every scheduling decision they make. Betting against the house, over time, is a sure way to lose money.

Why aren't EHRs equipped for scheduling in a more optimized way? Because, as we've said before, the math required for such processes goes far beyond simple calculations in Excel. A big reason for the mathematical complexities in asset optimization is the size of the solution space from which an optimal solution must be drawn. Consider a thirty-chair infusion center that completes sixty to seventy treatments per day, of four different duration lengths (e.g., one hour, two to three hours, four to five hours, and six-plus hours). The number of unique ways in which the calendar of appointments could be built for a single day is an integer with *more than a hundred zeros behind it.*

In a similar way, consider a single provider with eight available clinic hours per day who sees four different types of patients (e.g., new patients, returning routine patients, returning complex cases, and postprocedure follow-ups), each of which requires a different duration of one-on-one time with the provider. There are over five thousand trillion different ways of designing an appointment template for that provider for a single day. Even if one were to reject 99.9 percent of those templates as infeasible, that still leaves five trillion possible templates to consider. Can you see why the calendar and Excel spreadsheet methods that are currently employed in most health systems are so inadequate in solving the problem?

In surgical scheduling, global averaging is often used to estimate case durations. Averaging a one-level spine fusion with a multilevel fusion to come up with the "average time length of a spine case" is like putting your head in the freezer and your feet in the oven and averaging the two temperatures to come up with your "average body temperature." The answer would be neither accurate nor useful. There is a much more sophisticated way of looking at specific case types—along with the notes for those procedures—and coming up with data-science models that are far more useful and accurate in estimating the time-lengths of two very different types of spine case.

EHRs gloss over the shortcomings in their mathematical foundation by offering automated reports, dashboards, and limited forecasting features within their products. Although these features can be helpful to the frontline staff in analyzing what happened last week and avoiding some potential potholes in the future, they do not come close to approaching true mathematical sophistication in the form of constraint-based optimization algorithms, data-science-driven prediction models, simulation methods, and machine learning—all of which are required to truly optimize the operational performance of health systems.

4. EHRS DON'T INCORPORATE DOMAIN KNOWLEDGE ABOUT OPERATIONS

Like any good software tool, EHRs and their reporting modules can be configured and customized. You can use them to create and look at data, define metrics the way you want, and generate hundreds of reports. You can create various informational "snapshots" and choose a myriad of views.

As we will see in the next section of the book, however, the fundamental problem is that these metrics and reports are often the wrong ones to look at. The metrics that matter in matching demand and supply are unique for each asset and require a deep and sophisticated understanding of the operations and demand-supply patterns of that particular asset.

For example, in operating rooms, surgeons and service lines receive "block time"—it is the most precious resource an OR has to offer. As Chapter 4 explains, however, hanging your hat on "block utilization" as the metric you use to rightsize block allocation is highly misguided. It measures, rewards, and penalizes elements that have nothing to do with efficient use of OR time, as we will see.

Displaying numbers and reports without deep operational context hurts the process more than helps it.

Measuring "chair turns" for infusion chairs is equally meaningless. If your infusion center has a patient mix biased toward shorter infusions, and my infusion center serves sicker patients that require longer regimens, what sense does it make to compare the number of times our chairs turn? That's a bit like comparing table turnover at a fast-food restaurant to that of a fine-dining establishment.

Displaying numbers and reports without deep operational context hurts the process more than helps it.

OTHER TOOLS AND WHY THEY DON'T WORK

Next, let's look at the dashboards and reporting tools that are often built *on top* of data from the EHR. Much as with EHR reports, most of these tools only "describe," or at best, "diagnose," the problem rather than provide predictions or prescriptions to actually solve it.

Below are brief analyses of the descriptive and diagnostic analytics these tools typically provide, as compared to the more useful predictive and prescriptive analytics that a more mathematically robust tool can offer.

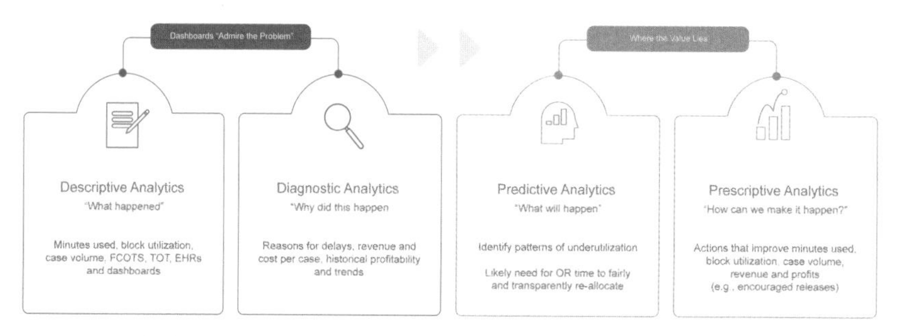

Analytics that predict and prescribe actions are far more useful than those that just describe the past: Perioperative example

Descriptive Analytics—similar to weighing machines and thermometers: Before using a weighing machine or a thermometer, most people have a defined hypothesis they want to confirm. The answer they receive from these tools either confirms or denies their hypothesis. But it does not tell them the underlying cause for the output, nor does it give them a sense of how to move forward. Similarly, in an OR, knowing your historical case volumes, times, turnovers, delays—the type of descriptive data provided by EHRs and other dashboards and tools—merely underlines the issues you probably already know you have.

Diagnostic Analytics (e.g., "drill-down dashboards"): Once you look at your thermometer and see you have a high fever, perhaps you can surmise that "I have a fever because I was out in the rain for a couple of hours and got soaking wet." That is certainly progress, but it still doesn't help cure your fever. In a similar way, all the slicing, dicing, and deep diving that diagnostic dashboards provide are like explanations of yesterday's weather. Wonderful for understanding where and when last night's storm occurred, but still not helpful in deciding whether to take an umbrella with you when you go out *today*.

In the OR world, diagnostic analytics might help you understand that Dr. Jones's cases always run late because he habitually requests too little time and doesn't think in terms of "wheels-in to wheels-out." However, they still don't *solve* anything unless and until they can help Dr. Jones learn to request the right amount of time in the future and give him an easy means to do so.

Predictive Analytics—think Google Maps: The real power of analytics starts to come into play when we gain the ability to forecast meaningful future events. Think about Google Maps and how it predicts that it's going to take you fifty-four minutes to get home from the airport three days from now after you land at SFO at 5:00 p.m. That's information that can help you plan the rest of your evening. How does Google Maps work such magic? By mining historical data from millions of trips drivers have taken over the years—by day of week, segment of road, weather conditions, whether it's a public holiday or not, and a host of other factors. The app has no way of knowing exactly who will be on the road when you take your trip, nor can it rate their driving skills, but it can access a lot of data by which to model and predict your likely drive time from point A to point B. In a similar way, time-stamp data can be mined to help an OR scheduler predict which surgeons/block owners will not use their time well in the future or help an infusion-center manager conclude, "I'm going to have a chaotic day next Wednesday between 10:00 a.m. and 2:00 p.m., and I need to plan for it." Analytics really start to add value when they give you specific information about the future you can use to solve a potential problem.

Prescriptive Analytics—think bin packing at UPS/FedEx or rerouting by Waze: The most useful type of analytics is one that drives high-value actions—for example, when Waze tells you to take a different route to shorten the length of your journey. Or when FedEx and UPS use forecasting algorithms to predict the volume and mix of packages they will receive from all around the country and are able to put the right number of planes, trucks, and drivers in the right places at the right times to handle the stochastic demand. Examples in the healthcare world: a surgical department is able to alter its staffing patterns for a future day on which a surge in case volumes is predicted, or an infusion center is able to anticipate a chaotic Wednesday afternoon two weeks from now and shift a couple of patient appointments to better slots in order to "flatten the peak" and fit more patients in.

Many more such examples lie ahead in upcoming chapters.

Going forward, hospitals will need better tools, with better math and more predictive and prescriptive capability than EHR reports and dashboards can provide. Operational teams will need to go beyond describing or diagnosing problems to actually predicting what's likely to happen and making action adjustments in anticipation—as illustrated by Waze, Uber surge pricing, and so many other real-world examples we all encounter in our day-to-day lives.

***

So … we have explored the financial and operational challenges that lie ahead for healthcare, and we have made the case that optimizing asset utilization is no longer a luxury but a necessity. We have shared our belief that—as with the ridesharing and airline industries we've looked at—it is those "early movers" who recognize the value of optimization that will gain the financial traction and the market share.

And we hope we have also made the case that the current scheduling system, based on calendars and block grids, is simply not designed for addressing the sophisticated challenges of asset optimization.

In the next section of the book, we will look at some specific assets within the healthcare industry, such as operating rooms, inpatient beds, infusion centers, clinics, and emergency departments. We will discuss some of the challenges inherent in trying to effectively utilize these assets. We'll analyze why current approaches aren't working and explore what the features and mechanics of a better system might look like. By reading Part Two, we believe you will gain a clearer understanding of the system-wide problem of suboptimal asset management as well as better insight into the nature of the solution.

PART TWO

CHAPTER 4

OPERATING ROOMS—NO BLOCK LEFT BEHIND

Ask any OR manager or VP of surgery, and they will tell you—as they have told us—that one of the biggest challenges they face is the efficient use of OR time. OR managers struggle to balance revenue and utilization targets with surgeon and staff preferences, all against a backdrop of supply-and-demand considerations that are tricky to predict and balance.

Operating rooms are vitally important to health systems because they are the economic backbone of a hospital, often generating more than 50 percent of revenues and profits for the institution. A single block of OR time (approximately five hundred minutes) can generate $50,000 to $100,000 or more in revenue per day, depending on the payer and case mix. ORs are also the most expensive resource to operate. So when it comes to allocating OR time, "Every block is sacred."

Yet the way blocks of OR time are allocated, taken away, released into open time, and requested by other surgeons leaves a great deal of room for optimization.

WHY DO "OR BLOCKS" EXIST?

As most people involved in OR scheduling know, the concept of awarding surgeons, service lines, or surgery groups half days or full days of "blocks" of contiguous time was created for the purposes of efficiency.

As we have discussed before, *all* medical appointments require the right mix of supply elements—personnel, room, equipment, etc.—to come together at the right time. But the operating room is especially complex in this regard. In order for a surgery to take place, you need the patient; the *right* surgical staff, including (but not limited to) surgeon(s), anesthesiologists, nurses, and surgical technologists; the room with the table; all the correct supplies; and all the needed equipment—for example, a specialized robot or electrosurgical tool. If you're doing a knee or hip replacement, you need the "replacement parts," etc.

The supply elements required in the OR depend heavily on the nature of the surgery. An orthopedic surgeon, a neurosurgeon, and a cardiac surgeon need very different and specialized sets of instruments, soft goods, rooms, technologies, and staff. In addition, every surgeon's available time is limited, based on a complex schedule of other professional commitments.

Imagine you're a scheduler trying to make the best use of ten operating rooms open between 7:00 a.m. and 5:00 p.m. Now let's say you need to schedule four cardiac surgeries for a particular surgeon or service line over the course of a particular day. Considering all the elements that need to come together for a cardiac surgery (versus, say, an orthopedic surgery), it's simply more efficient to use the same OR for all four cardiac surgeries, rather than moving all the surgical elements around from OR to OR with each new operation.

This has evolved into a system whereby the hospital sets aside a *block* of time and allots it to a surgeon or service line or group so that it becomes easy for that surgeon to conduct multiple similar surgeries in sequence with the same team, room, and equipment. These blocks of time are then reserved on a recurring basis for that surgeon or service line.

Surgeons typically have to organize their lives around clinic days, teaching, and other commitments, so they have many restrictions on which days they can avail themselves of block time. Thus, they naturally become somewhat territorial about their OR time and do not want to lose the precious blocks they have been granted. There is also an element of seniority and status when it comes to "who is entitled to what time"—if Tiger Woods has been assigned the 8:00 a.m. tee time on Wednesdays at Pebble Beach, the club will naturally be hesitant to reassign that tee time to another golfer.

So hospitals tend to assign block time based on the preferences, constraints, and availability of surgeons, sometimes taking into account their seniority and status. They also use block time as an incentive for attracting new surgical talent. Understandably, there is constant pressure from the surgical side for more block time. But at the same time, hospital administrators must ensure that their most valuable asset, the OR suite, is being utilized as fully as possible, while also leaving an appropriate amount of "open posting time" (OR time that can be claimed on a first-come-first-served basis) available so that emergent surgeries, or those to be done by surgeons without block time, can be performed as needed.

In most hospitals, surgeons are encouraged to release upcoming blocks of time they know they will not be using due to planned absences such as vacations or attendance at conferences, so that other surgeons may use them. If a surgeon has not scheduled any surgeries for their assigned block by, say, a week before the date, the hospital's system may "auto release" their block to others.

Hospitals must assess and reassess the utilization of their OR blocks on a continual basis. If one surgeon is routinely underutilizing their blocks and another surgeon routinely needs more OR time, blocks may need to be reassigned. This process, however, can be contentious and somewhat political. For one thing, calculating block utilization is a cumbersome, time-consuming process and subject to a variety of "rules" not everyone agrees with. For another, there is often disagreement over the underlying data itself. There may also be disagreement as to who exactly is responsible for that "first-case delay" or "turnover time," which counts against the block owner in such calculations. So what you are left with is a not-so-useful metric being used to allocate one of the most expensive resources in a hospital!

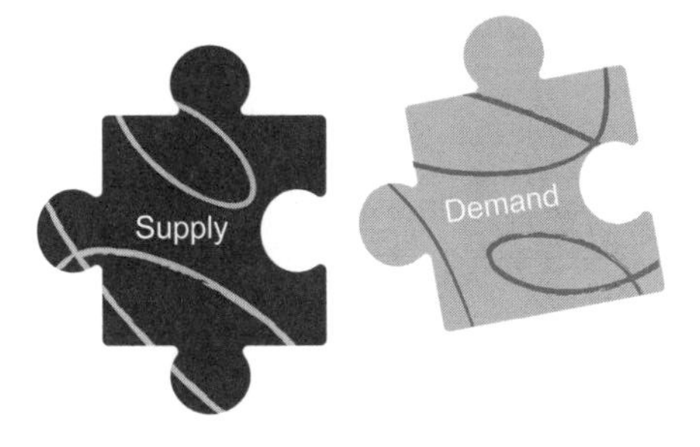

MATCHING SUPPLY AND DEMAND IN THE OR

As is true for all of the various healthcare assets we will be examining in this section of the book, supply and demand are the nub of the OR scheduling challenge. Matching supply and demand in the OR arena is notoriously thorny. The fundamental issues are as follows:

OR SUPPLY

Supply of block time—and OR time in general—is limited by the number of rooms, the equipment available, and the availability of staff (nurses, techs, anesthesiologists). It is also severely limited by the number of blocks that have already been allotted. The traditional method of allocating block time leads to a situation whereby a number of health systems we have worked with find themselves "heavily blocked," with not much open time left to attract new volume.

OR DEMAND

Demand, however, is fluid, unpredictable, and volatile, as in other healthcare areas. Just because a surgeon "owns" a reserved block of time, that does not mean he or she has more predictable scheduling needs than a surgeon who does not own block time. All surgeons require a varying number of minutes of OR time from week to week, due to a variety of reasons that are impossible to predict, such as the number of patients they see in the clinic each week, the percentage of those patients that will require a surgical procedure, and the lead time needed for scheduling each procedure.

This graphic gives you a sense of the level of volatility involved in block usage:

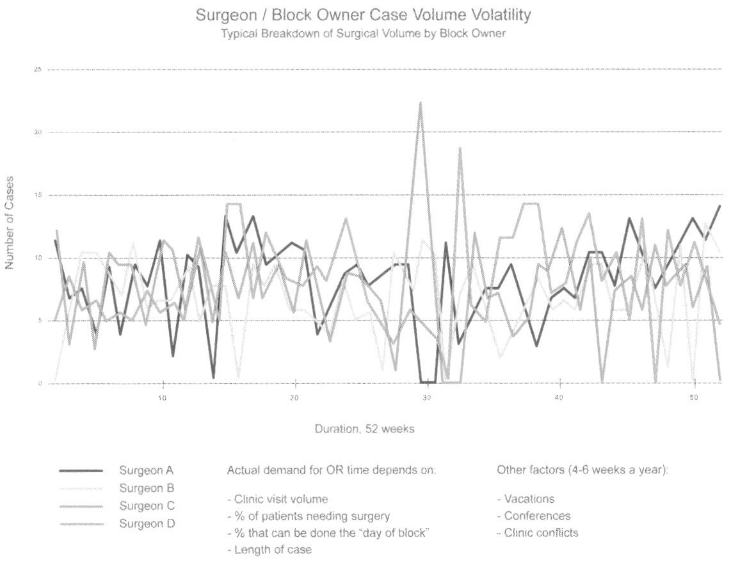

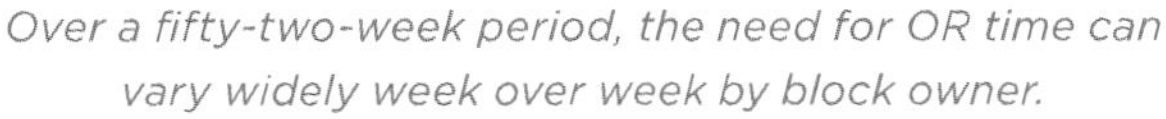

Over a fifty-two-week period, the need for OR time can vary widely week over week by block owner.

THE FLAWED WAY OR BLOCKS ARE CURRENTLY MANAGED

Yet most health systems assign blocks of OR time in the relatively static manner described above (i.e., Dr. X gets a full-day block on Monday and Thursday, or Orthopedics gets two blocks every day of the week). Decisions about block time are often based on historical rules for how blocks have been allotted and distributed in the past. Changes to the OR's block schedule tend to be slow and are usually made by an OR committee that meets monthly or quarterly.

To make matters worse, most health systems use a flawed metric, *block utilization*, to make decisions about the number of blocks that should be awarded or taken away from surgeons and service lines. Block utilization is a gross average that penalizes surgeons for small, fragmented chunks of unused time (room turnover, finishing early, late first-case starts, etc.) that are often beyond the surgeon's control and cannot be usefully repurposed into booking additional cases.

The voluntary process most health systems use for releasing blocks of time surgeons no longer need and for requesting new blocks of time tends to be manual, labor intensive, and error prone.

Furthermore, the block utilization metric does not account for the inherent volatility in case durations across surgeons and service lines. Anytime a surgeon "opens up" a seventy-five-year-old patient with multiple preexisting conditions, there is no telling what kind of complications they are going to run into or how long that procedure is going to take. This volatility makes any relative comparison of block utilization numbers inherently flawed.

The voluntary process most health systems use for releasing

blocks of time surgeons no longer need and for requesting new blocks of time tends to be manual, labor intensive, and error prone. It typically involves phone calls, text messages, emails, voice mails, faxes, Post-it notes, and drop-by visits to consummate the "transactions." It is old world, all the way. In fact, many of the clinics we meet with are not even aware they are expected to release time proactively.

The nonvoluntary "auto-release" feature mentioned above—whereby hospitals release blocks of time that have not been scheduled for procedures by the block owner—is a blunt instrument that simply doesn't work very well. That is because the natural lead time for scheduling a procedure varies widely across surgeons and service lines. By the time the hospital auto-releases a block, it is often too late for other surgeons to realistically schedule procedures into those times.

The net result of all of the above is that blocks of OR time are often left unused, even as surgeons who would have valued the opportunity to schedule some additional cases are forced to push those cases out to subsequent days or weeks, or to add them on at the end of the day. This can create overtime, leading to significantly higher costs and patient/staff dissatisfaction.

"Let's face it, block allocation processes have historically not been very efficient or data driven," says Brian Dawson, VP of Perioperative Services, CommonSpirit Health, one of the largest health systems in the country. "In many ways, it's similar to how Congress passes bills—surgeons lobby for more time, we spend hours pulling data and proposing changes, and blocks get debated and resolved in the OR committee as best we can."

LINKING INDIVIDUAL SERVICES

In addition to supply-demand *matching* issues, there are some common issues around the *linking* of services in the OR. These can occur both "upstream" and "downstream"—before and after the surgical procedure.

Upstream: As discussed above, one of the challenges ORs face in planning for the week or month ahead is the variability in the time each provider will actually need to use within their block. Setbacks in patient readiness are another potential upstream issue. Before going into surgery—on the "day of"—each patient must have completed the required insurance authorizations and the preop routine enabling them to undergo surgery. In addition, trauma cases and other emergent situations sometimes occur, forcing surgeries to be "bumped." This can create unexpected changes to the schedule.

Downstream: Postsurgery, some patients will need to be taken to the ICU (intensive care unit) for recovery, others to the PACU (post-anesthesia care unit). The capacity available in these key areas can create a major bottleneck as to when the current patient can be wheeled out and the next patient wheeled in. Similarly, when moving patients *from* the PACU, the availability, or lack thereof, of beds in the right downstream unit can also hold things up, as we will see in the next chapter.

In an ideal world, we would have visibility into the end-to-end dynamics of the system. We would be able to predict a patient's need for surgical and follow-up care, starting from their first clinic appointment and continuing through their discharge from the hospital. Doing this for each patient turns out to be quite difficult—a bit like anticipating exactly which driver will execute a lane change at any specific moment in time. However, it *is* possible, in aggregate, to predict *patterns* of need for OR time, make that time available, and get more cases done in prime time. Let's look at how next.

SOLVING THE OR PROBLEM

Clearly, health systems *want* to solve the OR problem, but they often have trouble accessing performance data in a timely and convenient manner and using that data in an effective way. What typically happens is that reports of block and room utilization, on-time starts, turnovers, and other metrics are extracted from the EHR, assembled into PDFs, and distributed to surgeons and service lines via email. The definition and purpose of the key metrics being displayed is often unclear, and there are frequent disagreements about the accuracy and meaning of the numbers. As a result, the reports usually fail to drive actionable change.

"As a perioperative leader, I oversaw a large team of physicians, nurses, managers, and IT specialists to analyze the operational statistics, develop dashboards, and implement reports; this process demanded a large manpower commitment," says Dr. Jose Melendez, chief medical officer at UCHealth's South Region. "Like many organizations, we have homegrown queries on top of our electronic medical record (EMR). Before using mathematical optimization solutions, my team

produced tons of data, requiring immense time commitments to pull it together into actionable reports, and at the end of the day, I was looking backward. I needed something that was less time-consuming and more forward-looking."

Solving the problem of suboptimal OR utilization requires a learning system and process that

- improves OR access for all surgeons and patients;
- improves asset-utilization rates for hospitals;
- creates greater accountability for block owners; and
- creates transparency and visibility into metrics *that matter* and that everyone can agree on.

The engine needed for this solution involves a combination of 1) prescriptive analytics that provide actionable recommendations, and 2) scalable software tools that enable easier planning, decision-making, block rightsizing, and the freeing up of more open time.

THE CORNERSTONES OF A BETTER SYSTEM

We believe that the cornerstones of a system to accomplish the above requirements include the following:

1. GET MORE TIME RELEASED EARLIER AND MAKE FINDING OPEN TIME AS EASY AS USING OPENTABLE

Currently, 15 to 25 percent of cases are done outside of block time, while 10 to 15 percent of block time goes underutilized or abandoned in many hospitals[22], owing to a multitude of reasons—a surgeon attending a conference who forgets to cancel her block, a patient canceling a procedure at a late date, a surgeon holding on to block time, hoping unsuccessfully to fill it at the last minute, and more.

Many hospitals, however, have begun to use predictive analytical tools to address the problem and are achieving positive results. As of today, at over fifteen hundred ORs[23] across the country, algorithms are being employed to monitor booking patterns to identify blocks that are *likely to be underutilized* and to remind surgeons and schedulers to proactively release these blocks into an open pool their colleagues can use. This "simple" step alone facilitates more time being released well in advance of the block, which greatly increases the likelihood of other surgeons being able to see, pick up, and use that time.

Duke University Health System is a great example:

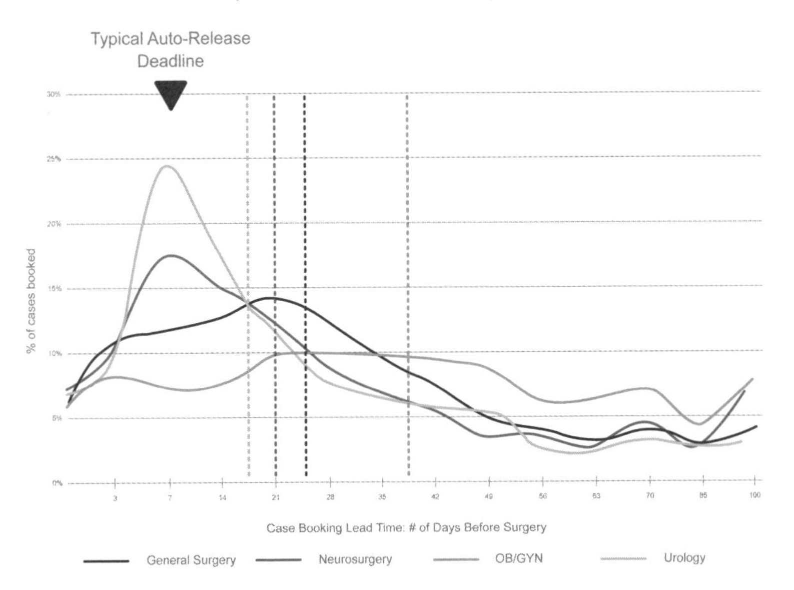

Booking patterns vary widely by surgeon and clinic, yet "auto-release" deadlines are usually fixed.

22 LeanTaaS internal customer data.

23 Ibid.

"In the past, surgeons would often tell their colleagues in the same service line they weren't going to use their block time in case their colleagues wanted it, but they wouldn't tell the schedulers or managers," says Wendy Webster, director of clinical operations, departments of Surgery and Neurosurgery, at Duke University Medical Center. "Because of this, block time went unused at a time when some service lines were trying to grow at a more rapid rate than others and needed more time. Creating a more transparent release process with easy-to-use 'nudges' has been a big win for our surgeons. Over 2.5 million minutes were released in our 105 ORs at Duke using these Release Reminders, and weeks before they normally would be available. Then making the right time available to the right clinics and surgeons, taking into account their need for specialized staffing, equipment, and room types, has led to a lot of efficiency—we are able to do eight more cases per day in our ORs as a result."

Parkview Medical Center is another example. It shows that similar benefits can be realized in community hospitals:

"Online tools that have made access to open time occur sooner have made the surgeons happy, and we're getting patients in faster. Adopting these tools has increased surgeon satisfaction, increased patient satisfaction, and has made the whole scheduling process outside of normal block time smoother," says Dr. James Caldwell, Vice President of Medical Surgical Operations & Medical Director of Surgical Services

> *at Parkview. "It has fixed so many problems and streamlined our systems so much."*

To gain similar benefits in your own organization is largely a matter of mindset and willingness to adapt to new tools. Think about your own booking behavior as a customer. When was the last time you called a travel agent to book a flight instead of using online tools such as Priceline or Travelocity? The last time you called a restaurant to make a dinner reservation instead of using OpenTable?

So why are we still telephoning and emailing—even faxing—back and forth between clinics and ORs to look at open times of choice? Improved math and sophisticated, cloud-based tools already exist to make the process more efficient, accessible, and transparent. And the results are measurable.

Why are we still telephoning and emailing—even faxing—back and forth between clinics and ORs to look at open times of choice? Improved math and sophisticated, cloud-based tools already exist to make the process more efficient, accessible, and transparent. And the results are measurable.

2. DON'T RELY ON "TRIBAL" RULES FOR BOOKING

Many organizations fall back on historical and tribal "rules" to decide how much time to block and how much to keep open in their ORs. In working with over 150 hospitals, we have heard many of these rules—for example, the "80/20" (sometimes "90/10") rule. It refers to the amount of block time versus open time believed to be ideal. Everyone in healthcare knows they need to keep *some* OR time open in order to handle unexpected volume from emergency rooms and late add-ons, so they often use such predetermined ratios. If they're running a Level 1 trauma center, their ratio might include more open time, but it's still a preset ratio.

Sometimes the rules are based on what an influential surgeon wants or doesn't want. Sometimes they may be shaped by a consulting study or research paper or supposed industry "benchmark." Using benchmarks in healthcare is a bit like borrowing someone else's fingerprint—it doesn't make sense, because *your* case mix, *your* patient population, *your* surgeon population, and *your* facility's characteristics are all unique to you. One-size-fits-all rules are not only mathematically imprecise, but they also lead to poor decisions. Imagine if the Waze app, despite having access to massive amounts of historical data on driving speeds per each segment of road for every minute of the day, failed to use that data but instead relied on a generic rule such as, "If it's rush hour on the highway, set the speed at twenty miles an hour." How accurate would Waze be at predicting the best routes for drivers?

There are better ways to look at the data. In order to be more like Waze, you should be using your data to answer meaningful questions such as the following:

- How much of our case volume by day of week and time of year is truly elective versus urgent/emergent?
- Do we really have a 20 percent volume of add-on cases, or is this just our surgeons' way of squeezing patients in because they couldn't find any available time on the schedule? Really dig into the data—how could a case possibly be considered urgent-emergent if it was in the depot five days before being gridded?
- Does a particular service line with its own trauma room really need a dedicated OR every day?

- What is the lead time for booking cases for this very particular type of surgery?

All of this data exists, and if you can mine it and use it in the right way, you can come up with a far more accurate way of calculating the best mix of open versus block time in your ORs. You can then set your "thermostat" in the most efficient way to create that right mix for your particular facility.

3. EVALUATE BLOCK USAGE IN A MORE MEANINGFUL WAY

> *The smallest quantum of usable time in an OR is not a minute; it's the smallest amount of time needed to do a case.*
>
> *—Dr. Allan Kirk, surgeon-in-chief for the Duke University Health System*

Dr. Kirk's astute observation reveals a fundamental flaw in using "block utilization" as a metric to evaluate block owners. Specifically: When you lay out utilization patterns by surgeon, there are holes, or white spaces, in them. However, not all "white spaces" are created equal. A ten-minute first-case delay, a fifteen-minute turnover delay, and a case length overestimated by thirty minutes cannot be brought together to create enough open time for most ORs to fit a case in.

When EHRs or other tools define block utilization, they typically fail to differentiate between "meaningful holes" and "small grains of sand" that may matter from a day-to-day operational perspective but don't really matter from an *access* perspective. What's worse is that this metric tends to focus surgeons on those small delays in their surgical process that they are not responsible for and to penalize them for the wrong things. For example, surgeons finishing cases early are often penalized for *efficiency*, resulting in a direct misalignment of incentives. This is why conversations centered around block utilization often end with no actionable result or decision.

The nature of surgical case time-lengths is stochastic—no one can predict a priori, beyond a certain range, how long an OR procedure is going to take. And depending on the department/case type/patient population, the "standard deviation" from the mean or median may be large or small. Sports orthopedics may have a relatively "tight" variation; neurosurgery times may be quite varied. This leads to the second drawback in using block utilization as the metric: it takes no such variability into account.

The magic in making block rightsizing decisions is to identify block owners that are frequently leaving large enough holes unused that you can take a block or more away from them and still allow them to do all of their procedures—we call this "Collectable Time."

Collectable Time, as a metric, is far more surgeon-centric and actionable than mere block utilization, and it is one that block owners actually understand and appreciate because it levels the playing field between those with high variability in case lengths and those with more predictable variability.

For example, see the following graphic. Two surgeons with the same *total* unused time (and hence the same block utilization), but with widely different distribution of that unused time, have vastly different Collectable Time. Surgeon A has a lot of Collectable Time, but Surgeon B, who has much more unpredictable case durations, has very little, even though they both have exactly the same block utilization.

Two surgeons with the exact same block utilization can leave vastly different amounts of "Usable/Collectable Time."

Collectable Time also doesn't penalize surgeons for delays that are often beyond their control—paperwork problems, staff delays, etc.—or for using time more efficiently than anticipated. Often a surgeon, through no fault of their own, ends up leaving a little time each day, which is smaller than the length of a case the OR can accommodate. Block utilization would penalize them for that amount, but Collectable Time ignores it.

University Hospital at The Ohio State University Wexner Medical Center has adopted this new way of understanding block usage and has seen positive results.

> *"Focusing on just the large chunks of unused block time has helped facilitate a lot more surgeon accountability for their allocated block time," says Dr. Laura Phieffer, medical director of the above facility. "This has helped drive actionable data-driven discussions with surgeons and decision-making in our Block Committee meetings."*

Not all "white spaces" are usable.

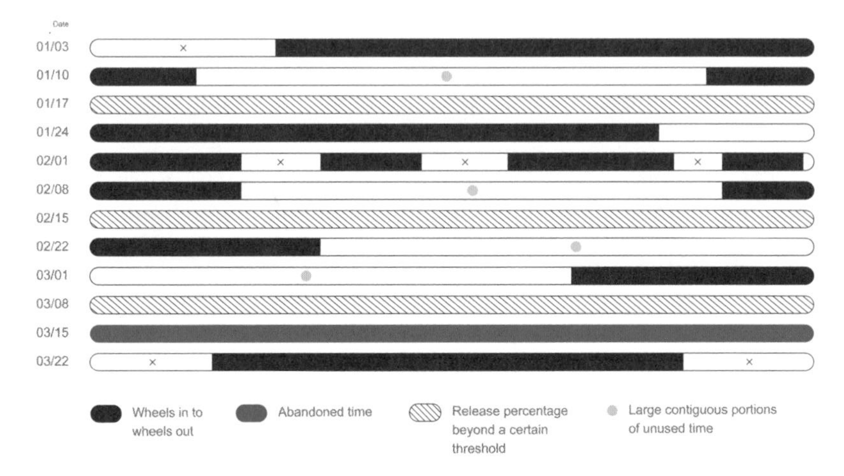

4. PROVIDE CLEAN, CREDIBLE, AND TRANSPARENT REAL-TIME DATA

We live in a world where we expect to be able to quickly pull up bank balances on our mobile device anytime, twenty-four seven, and we expect them to be accurate and current. We expect to be able to view, in real time, current stock prices, recent credit card purchases, and the exact location of a package in transit—and we expect this information to be accurate and presented in visually clear, easily understandable ways.

By contrast, much of the important information used in OR scheduling is opaque, difficult to access, and disputable. In the OR world, we have seen the following:

- Multihospital systems in which each location uses its own definition of turnover time, block utilization, prime time, first-case delays, and more.
- An abundance of reports but little action taken. In many systems, dozens, or even hundreds, of reports are generated every month—paper-based, EHR-based, Tableau- and Excel-based. Recipients often don't know what to do with the data, so they request even more data in order to gain clarity. The

number of meetings that end with "Let's get someone to run that report" is astonishing. Why isn't that "report" available at everyone's fingertips, like bank account information or Amazon tracking reports, so it can be discussed in the present moment and action steps can be taken?

- Many questions, few answers. Almost every hospital we have worked with has people questioning the numbers. Perhaps it's a surgeon who doesn't believe their utilization data because they thought they released a block and don't see credit for it, or is confused because they don't have an easy way to visualize why they are "inefficient when they flip rooms." Perhaps there is disagreement over whether case lengths are "wheels-in to wheels-out" or "surgeon-in to surgeon-out." Or someone doesn't believe a metric is well-defined or reported correctly. Often one report doesn't "foot," or agree, with another. These multiple sources of truth end up creating a lack of faith in the data, which results in the data being ignored.

Reporting should not hinge on belief, opinion, or point of view. But that's how things work when you use nonscalable, person-dependent ways of gathering data.

Reporting should not hinge on belief, opinion, or point of view. But that's how things work when you use nonscalable, person-dependent ways of gathering data, and when you fail to establish a shared understanding of what the data means, which metrics are important and actionable (and why), and which definition of each metric everyone will converge upon.

If your data analyst or report writer takes a day off to care for a sick child and you can't make a decision on a resource that is worth a great deal to you, that's a fairly significant problem.

Instead, imagine a world where you could access your report of choice from a mobile device twenty-four seven without needing anyone to create it for you, where the data was clean and displayed in a visually intuitive form, where the data was updated consistently in a timely manner, where the metrics were unambiguously defined and curated to optimize actionability and decision-making, and where you could drill down into each metric with the tap of a screen. That's the world we live in outside the OR, and we can replicate that world in the OR arena.

5. ENGAGE STAKEHOLDERS IN THE DATA

While we are playing the imagination game, let's also picture a world in which perioperative leaders received a succinct mobile text each week to let them know what's going on in their ORs. For example, surgeons would receive a short and meaningful summary of their OR activity with a link that takes them to better understanding of

- how they are contributing to OR volume;
- how their performance measures are trending;
- ways to improve their utilization; and
- recurring causes of delays when they operate.

The right data should always be at your fingertips and proactively "pushed" to the right person.

Imagine, too, if every surgical committee had a browser in the room and was able to navigate to answers to key questions *while the meeting was still going on*: "Are we using our robots efficiently?" "Which ten surgeons need the most help with their case-length predictions?" "Which service lines are dropping the most in volume, and is that unusual at this time of year?"

What if, in fact, we were able to evolve past the "report" mindset entirely? What if, instead of waiting on reports, everyone had the self-service capability to access the important data in real time? Take it one step further. What if that data was believable and could be shown to have clear and demonstrable outcomes—much in the same way that a college website shows how a midterm grade affects the final grade? What if we could send "early warnings" to block owners to show that their performance was trending in the wrong direction and might result in the loss of OR time or that "on Wednesdays when you work with Case Team B, you always start late."

That world can and does exist, as Dio Sumagaysay of Oregon Health & Science University attests:

> *"We have decided to open up web-based reporting tools to every surgeon and service line across the board," he says, "so there is full transparency on the key metrics we care about across the institution. The process of cleaning up the data, agreeing on the metrics, and making [the information] universally accessible and useful has been game-changing culturally."*

6. DON'T USE "BAD MATH"

To extend an analogy we used earlier: If my head is in the freezer and my feet are in the oven, my body temperature, on average, could be a healthy 98.6 degrees—but I might not be alive to care. That's an example of how math can be misleading if not used intelligently.

- Think about many of the metrics OR decision-makers use that are based on averages—auto-release triggers that open up time if cases are not scheduled by a certain date, average turnover times, average first-case delay times, average block utilization. What meaningful information do these averages really give us?
- Think about the error that is created by using "global averaging" to estimate case lengths—especially in complex surgical cases like multilevel spine fusions, as we mentioned

earlier. What sense does it make to average a two-level fusion with other multilevel fusions? The number we produce tells us nothing about how long a specific fusion case will take.

Looking at data correctly is vital. Using sophisticated *classification* algorithms to identify the right types of cases—those that share meaningful characteristics—and then *clustering* those cases together to distill their means/medians/percentiles and other useful data makes eminent sense, but the hard work is in the "classification and clustering" methods. That's how Amazon's and Netflix's personalized recommendation engines work. The same type of thing can be done in the OR by putting the right math into service.

7. STOP "ADMIRING THE PROBLEM"

We expect airlines to text us if our flights are going to be delayed. We expect our credit cards to ping us if our card is being used in an unexpected locale. We expect Amazon to email us if a delivery estimate has changed. All of this is useful information.

But much of what passes for useful information in the OR setting flows from a backward-looking dashboard. Looking at what went wrong last quarter is a bit like raising an alarm after the horse has wandered out of the barn. We end up admiring the problem rather than solving it. But now, there is a tremendous amount of data available we can harness to usefully predict future outcomes and prescribe courses of action, such as the following:

Looking at what went wrong last quarter is a bit like raising an alarm after the horse has wandered out of the barn. We end up admiring the problem rather than solving it.

- Which blocks will likely go underutilized and should be released by the block owner;

- Which block owners have too much time and which ones have too little;
- Which surgeons' or teams' performance is trending in a statistically significant direction, either positive or negative;
- Whether the OR will have an unusually high volume or low volume day on day X and whether staffing should be adjusted accordingly;
- The likeliest range of case-length times, based on other cases similar to this one.

And so on. When we predict these kinds of important answers and prescribe actions against them, we create tangible, measurable value in our ORs. Let's stop admiring the problem; let's solve it. We have the technology.

Closely tied to ORs is another major asset that hospitals must manage with great efficiency: inpatient beds. As we'll see in the next chapter, this asset carries its own unique set of challenges.

CHAPTER 5

INPATIENT BEDS—RIGHT PATIENT, RIGHT BED, RIGHT TIME

Historically, inpatient bed capacity has represented a substantial chunk of potential revenue for a hospital, second only to the OR. As growth in *out*patient volume continues to rise, hospitals need to strive to be more efficient with their *in*patient bed utilization than ever before.

WHY BED MANAGEMENT MATTERS

Even beyond the pure economic value of each bed—which is substantial—the *flow* of patients into and out of inpatient beds is also vitally important for several reasons.

AVOIDING THE "STANDARD" DAILY CHAOS

A common problem in nearly every hospital is a lack of available beds when patients from the ORs or ED (emergency department) need them, or when patients are transferred in from another facility. This often leads to a clogged system that backs up other key areas of the hospital—the ED, the OR, the PACU (post-anesthesia care unit), and more. Patients can't be moved from those units until there are open beds to accommodate them. Patient-placement personnel, house supervisors, unit charge nurses, and even senior hospital leaders often become involved in managing this daily chaos as best they can.

IMPROVING PATIENT ACCESS

Patient access is inversely proportional to LOS (length of stay). The more time outgoing patients spend in their beds, the fewer beds are available for incoming patients. Hence, the impact of even a small change in LOS in either direction can be quite significant. Reducing the amount of unnecessary time spent occupying a bed for outgoing patients is thus a meaningful goal.

CHANGING THE LOSE-LOSE FORMULA

The frustration patients and providers experience when waiting a long time for a bed after emergency or surgical care and still sometimes ending up with the "wrong" bed (i.e., a bed in a unit without the appropriate level of specialized care) can be avoided by increasing the throughput of the system and by improving the matching of supply-demand signals.

MEETING REIMBURSEMENT REQUIREMENTS

That extra (and sometimes avoidable) night a patient occupies a bed—and those extra hours a patient waits for discharge—may not be reimbursable, so it is vital for bed flow to be fluid and for patients not to be languishing in beds unnecessarily.

***

Improving *flow* into and out of inpatient beds is good for all parties involved—patients, hospital staff, physicians, surgeons, and finance departments. Achieving reduced wait times in the ED, PACU, and other hospital departments frees up resources to attend to patients sooner. Problems with boarding and discharge are diminished, and there is a generally "saner" experience for everyone involved. When patients aren't waiting hours for a bed while sitting in the ED, their experience improves, and broader metrics including Press Ganey scores, leave-without-being-seen rates, and core clinical quality measures take a turn for the better.

When patients aren't waiting hours for a bed while sitting in the ED, their experience improves, and broader metrics including Press Ganey scores, leave-without-being-seen rates, and core clinical quality measures take a turn for the better.

The problem is that the traditional approach of "Let's just try to discharge patients faster" doesn't work—there are more complex and systemic sets of problems to address.

MATCHING SUPPLY AND DEMAND FOR INPATIENT BEDS

Inpatient beds create their own unique supply-demand challenges. Simply put, in most hospitals, the *demand* for beds on a daily basis usually swells before the *supply* of right beds opens up. (And of course, as in most other areas of the healthcare system, both demand and supply behave in an unpredictable way throughout the course of the day.)

A more "mathematical" way of stating the problem is that the *arrival patterns* of patients (from the OR, ED, cross-hospital transfers, and other areas) who need a bed is "out of phase" with the *departure patterns* of patients who are leaving the hospital, and this mismatch occurs in the wrong direction.

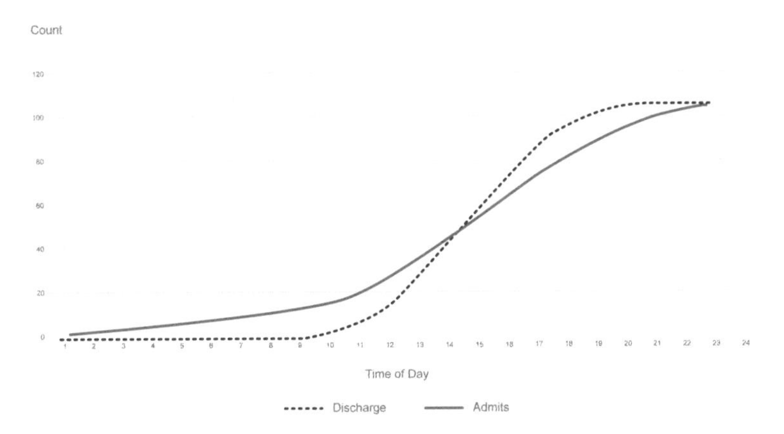

The daily demand for beds typically grows before beds are available.

As an example we can all relate to, hotel managers make this mismatch of arrival and departure patterns work in the *right* direction for themselves by establishing a checkout time of noon or earlier and a check-*in* time of 3:00 p.m. or later. Most hospitals, by contrast, experience the opposite situation: morning surgeries and overnight ED arrivals create a need for beds in the morning, while typical hospital discharge processes render beds available later in the day.

What makes this problem even more complex is that bed capacity must be divided into "units," as determined by the level of care and skilled staffing required. A patient recovering from serious heart surgery needs a very different level and/or type of specialized care from one who has had spinal or orthopedic surgery. So there is a "right" or "primary" unit to which a patient needs to be steered in order to access the right nursing care, equipment, and physicians.

All of the above add up to a limited availability of inpatient beds on a recurring basis.

THE DEMAND SIDE: WHO WILL SHOW UP WHEN, AND WHAT BED WILL THEY NEED?

To illustrate the demand side of the problem, let's think of the case of a hypothetical four-hundred-bed hospital—we'll call it Sunnyvale General—that's subdivided into twenty units of twenty beds each: ortho, neuro, cardio, relevant ICU units, and so on.

The need for beds at Sunnyvale General will be driven mainly by the following sources and circumstances:

When and how many elective, scheduled, and emergent surgeries of each type are done. The way in which OR blocks are allocated by day of week, and by surgeon or service line, will determine how many surgeries of each type will likely be done on a given day and therefore how much bed demand will be placed on each unit. For example, a heavy ortho surgery day will create a corresponding downstream demand for beds in that unit.

Volume and timing of add-on surgeries. Late-scheduled cases that surgeons add on outside of their block time, or in open time, may create the need for beds.

Inflow from the emergency department. Patients may come in from the ED requiring emergency surgery or other procedures that necessitate overnight stays.

Inflow from transfers. Some hospitals accept transfers from other institutions because they are able to provide more specialized care for certain conditions.

Patients who are admitted directly, planned or unplanned. These might be patients coming in for scheduled inpatient treatments or procedures, or they might be acutely ill patients who come in for an outpatient clinic visit and then need to be admitted.

So at Sunnyvale General, the "demand signal" from the above sources adds up in a different way on any given day and from week to week. Dealing with this stochasticity of demand is a constant problem Sunnyvale's staff faces in trying to place the right patient in the right bed at the right time.

THE SUPPLY SIDE: WHEN WILL A BED OPEN UP?

The *supply* of beds at Sunnyvale is driven by an equally complex set of dynamics that lead to variability in discharge volumes and bed-availability times.

Unit variances. The LOS of patients in each unit can vary widely. Some units may have a typical LOS as low as two days, others might have five-day average stays, and still others could have an even longer LOS. At Sunnydale General, Unit X, with a two-day LOS, will be freeing up, on average, ten of their twenty beds daily; Unit Y, with an LOS of five days, will be freeing up four beds per day, and so on.

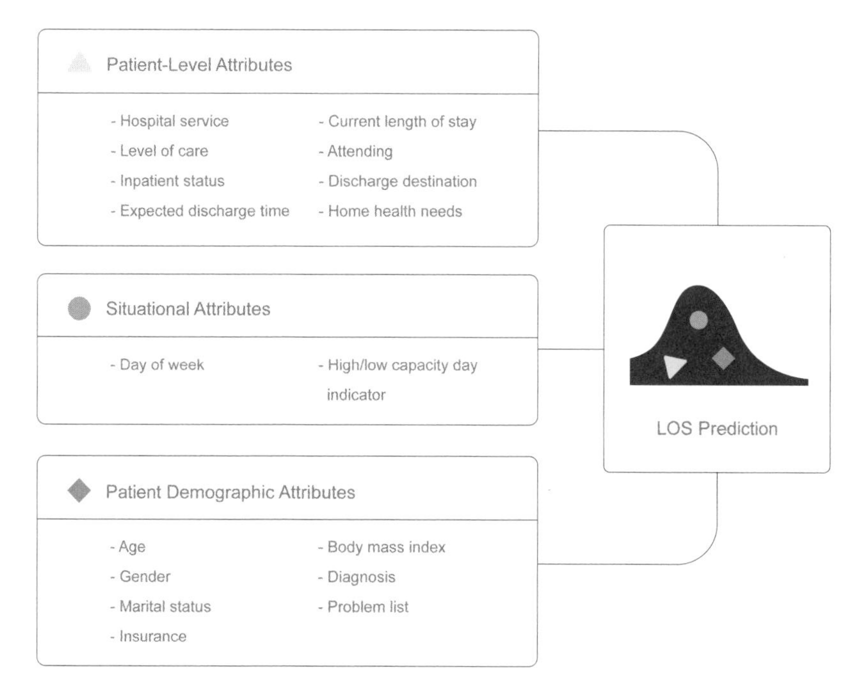

Factors that can help predict length of stay (LOS)

Discharge order times. In order to discharge a patient, the attending physician needs to write the discharge order, which can come at varying times of the day. As a rule of thumb, surgeons prefer to do their rounds in the morning before surgery starts so they can check on the recovery of their previous surgical patients, some of whom may be ready for discharge or transfer. There may be some patients who might need to pass a physical test (e.g., PT/OT) before discharge and others who may need a set of lab tests or a CT scan before the final discharge order can be written. These tests need to be cross-scheduled with other departments—adding another variable to the mix.

Other specialties (e.g., medicine) may do their rounds later in the morning due to teaching structures and admitting activities, which can often lead to delays in the writing of discharge orders, further exacerbated by patients' needs for testing and studies, as mentioned above.

Unpredictable delays in placing discharge orders. Even when a patient may be "nearly ready for discharge" pending one more test or result, there is no guarantee they will actually be discharged as planned. A certain percentage of patients who are expected to pass their PT/OT test or to have good lab results will not pass and may end up needing to stay an extra day or two.

Delays in discharge after the order is placed. Even after a patient has been cleared for discharge, multiple delays in making their bed available for an incoming patient can occur:

- "Avoidable delays": For example, problems with insurance documentation or with available space at a skilled nursing facility (SNF).
- Waiting for prescriptions: The discharge order might come with a set of prescriptions, and if the pharmacy is backed up, the patient may need to wait.
- Waiting for exit transport: Wheeling the patient from the bed to their family/transportation pickup spot requires appropriate support staff to be available in the right unit at the right time. It is not uncommon for patients to wait an hour or more simply to be wheeled to the exit.
- Waiting for transportation home: If the patient's family or other transportation isn't available at discharge time, that could further add to the delay.
- Waiting for medical equipment and supplies: Such items

(wheelchair, cane, wound care supplies) need to be ordered, picked up, and delivered to the patient. There may be delays in the availability of the items and/or the personnel to make these deliveries.

Waiting for housekeeping/environmental services: Once the patient leaves, getting the room and bed ready for the next patient requires the right services to be available in the right unit at the right time.

> *Effective inpatient capacity management improves patient care, provider experience, and hospital efficiency by strategically allocating a fixed supply of beds. This requires the ability to anticipate variances in bed demand across clinical specialties, patient demographic groups, and time.*
>
> *—Isobel Handler, Director, Operational Intelligence at UCHealth*

LINKING INDIVIDUAL SERVICES

As in ORs, there are service-linking issues related to inpatient beds that can occur both upstream and downstream.

Upstream: The number of patients who will be coming in from the ED and the units in which they will need beds cannot be predicted with a high degree of accuracy. Trying to predict when a particular surgery will be finished and when the patient will need a bed also has an "error bar" associated with it. And for those hospitals that accept transfers of patients from other facilities, there is uncertainty as to when a bed will be needed for a transfer patient.

Downstream: The ability to discharge patients depends on many downstream factors, some of which are beyond the hospital's control:

- Availability of a bed at an appropriate skilled nursing facility if that is the patient's destination.
- Transportation arrangements—either via the patient's family or via a professional transport service.
- Availability of results from lab / physical therapy / occupational therapy testing so the provider can clear the patient.
- Pharmacy services so the patient can be sent home with the medications they need.
- Procurement of durable medical equipment (DME) such as canes or walkers.

So every day, in thousands of hospitals across the country, patient-placement and supervisory staff do the best they can to plan ahead, unit by unit—but too often they have only rudimentary and insufficient tools at their disposal.

HOW SUPPLY AND DEMAND "MATCHING AND LINKING" ARE CURRENTLY HANDLED

Hospital personnel are well aware of the challenges of managing inpatient beds. So every day, in thousands of hospitals across the country, patient-placement and supervisory staff do the best they can to plan ahead, unit by unit—but too often they have only rudimentary and insufficient tools at their disposal.

- Nursing and staff huddles: These informal meetings start early in the morning and are repeated throughout the day. At each huddle, Excel or paper spreadsheets are reviewed to discuss and predict how many beds will open up and when, and to try to estimate demand for those beds by time of day.
- Deciding if and when to deploy "surge capacity" (i.e., capacity that the hospital is licensed for but that is opened up only when buffer capacity is needed).

Sometimes the day works out well, and most patients are placed in the right unit without prolonged waits. But sometimes patients are placed in the wrong unit and need to be transferred later, sometimes beds are lined up in the hallways of emergency rooms, and sometimes the most expensive beds in the hospital (ICUs, EDs, and PACUs) remain occupied all night by patients who no longer need them, leading to frustration, long waits, reduced reimbursements, and lower access to care.

TEN WAYS TO IMPROVE OPERATIONS

In our work across health systems, we have found ten potential levers that—if backed by the right predictive analytics engine, easy-to-use tools, and supporting people and processes—can make for much smoother operations in inpatient units. These levers address both present-moment issues and future-looking issues that fall into three categories: matching demand with supply, "unlocking" supply, and "shaping" demand.

1. PLACE PATIENTS USING SOPHISTICATED DEMAND-SUPPLY MODELS

The core approach to the optimal placement of patients is to accurately forecast and match demand and supply, unit by unit, by time of day, each and every day—given the existing structure of supply (e.g., the twenty units of Sunnyvale General, each with twenty beds) and the complex variety of demand signals.

Supply: The essential nugget of this approach on the supply side is to model the availability and timing of beds that will come available in each unit. Historical data can be used to mathematically create a "fingerprint": a model for each unit that predicts, for any given time of any given day of the week, the likely number of patients that will be discharged.

This model can be used to address incoming demand signals (see below) and enable hospital staff to make concrete placement decisions about individual patients; examples include the following:

- Wait for a bed in a specific unit because the right bed is likely to open up later in the day.
- Place the patient in an open (and right) bed now because it is available.
- Place the patient in the secondary or tertiary choice of units, instead of the primary one, because the right bed is unlikely to become free in the right unit within a reasonable amount of time.

Demand: Specifically tailored models can be used to create an "upcoming demand signal" at any time of the day for each element of demand:

- Volume from the OR: Elective surgeries typically have the most predictable demand signal. Using historical data, such as the expected wheels-in time and wheels-out time for this specific type of case, it is possible to estimate the time when each patient will need a bed.

- Volume from the ED: ED volume can be roughly predicted in finite segments of time (e.g., "Over the next hour we can expect eight to ten patients"). The conversion rate from ED to inpatient beds can also be modeled to estimate how many beds will likely be needed. This process becomes less precise as one tries to predict the need for beds in each specific unit, but even the macrosignal can be quite useful in gaining a sense of aggregate bed demand hour by hour.
- External transfers: This signal is typically hard to assess and is the least predictable, but fortunately, transfers are not usually a large source of demand for beds daily.
- Internal transfers: This signal is usually well known, based on the list of patients waiting to be placed in the right units after having been originally placed in the wrong unit.

These supply-and-demand side models can then be used to create tools that give patient-placement leaders visibility into upcoming demand and supply for beds, unit by unit, enabling them to make fact-based trade-offs. The software can also provide a "scratch pad" to allow decision-makers to experiment with potential bed distributions, much in the same way a chess player can play out various "what if" scenarios by using simulation tools. These tools can be quite powerful and can help placement personnel navigate their day, just as Waze helps you take the right turn at the right time to minimize the total time to your destination.

2. MAKE DATA-DRIVEN INTERNAL TRANSFER DECISIONS

The above demand-supply matching can be made even more effective if internal transfers are executed wisely. For example, by thinking a few moves ahead via the scenario-building "scratch pad" described above and moving the right patients to appropriate open beds, placement teams can open up the right slots to meet expected demand for high-value beds.

3. RIGHTSIZE YOUR UNIT CAPACITY (UNLOCK SUPPLY)

Fundamental unit sizing must be constructed correctly (i.e., using data from historical demand). If it is not, a higher percentage of patients will fail to be placed in their target units.

- This should be fairly intuitive. In Sunnyvale General, if a unit with twenty beds and a five-day LOS routinely needs to open between six and ten beds a day to meet its various sources of demand, it will frequently run out of space. Why? Because four patients, on average, are being discharged daily, while six to ten beds are needed.
- The correct number of required beds can be computed for a given unit by analyzing historical patient volumes by service and by day of week.
- After establishing theoretically optimal unit *sizes* by day of week, it is also important to consider unit *geography* and team assignments in order to ensure proper staffing and ease of rounding for attending physicians.
- In some cases, it might make sense to vary the size of certain units on a day-to-day basis by shifting some beds from one

unit to another based on higher expected demand in one unit than the other. Using data to make these trade-off decisions increases the likelihood that the units will be optimally sized on a given day.

4. LOOK HARD AT THE DEGREE OF SPECIALIZATION TO POOL CAPACITY WHERE POSSIBLE

If Sunnyvale has too many specialized units, this can lead to more cumbersome placement problems than if they created a larger pool of beds for patients to be placed in. If and when hospitals can combine several subunits and open a larger ("virtual") general medicine unit without sacrificing quality of care, they may be able to simplify patient placement and potentially co-locate patients assigned to a given clinical team.

5. SMOOTH THE PATIENT FLOW FROM THE OPERATING ROOM (OR)

The flow of patients from the OR into inpatient beds often accounts for 20 to 25 percent of the bed demand on any given day and can cause spikes in the inpatient census. The good news is that this flow is more "controllable" than the census contribution from the ED. Optimizing the elective surgery schedule with respect to recovery time can yield a flatter inpatient census.

Such "surgical smoothing" can be done by forecasting the volume and case mix of surgeries, along with the associated recovery time per case, and applying optimization techniques to build surgery templates that flatten the downstream inpatient census. This might entail recommending to OR committees that certain service lines/surgeons change the days on which they operate. The challenge here, of course, is that surgeons must be willing and able to comply with these changes, given their own clinic schedules. Such changes may be uncomfortable but may be worthwhile for those facilities really squeezed for inpatient capacity.

6. IMPROVE PROVIDER WORKFLOW

In many hospitals, physician rounding, particularly among medicine units, occurs in the late morning or early afternoon. This timing can lead to missed opportunities to discharge patients early in the day. Simple measures such as maintaining (when possible) a list of patients who can likely be discharged early and having providers round on these patients early in the morning can be hugely effective in opening beds earlier, provided clinical teams are able to cooperate.

Adding administrative support to provider teams to help plan and execute discharges can go a long way. One option is to schedule multidisciplinary rounds twice a day. Case managers can hold one session early in the morning to review discharges and a second session in late afternoon to review the roster and plan for discharges the following day or later. Also, dischargeable patients can be seen first (provided there are no patients with more acute needs) so that orders may be written earlier.

7. DON'T LET A DIME HOLD UP A DOLLAR—TAKE A HARD LOOK AT STAFFING LEVELS, TRANSPORTATION, AND HOURS OF OPERATIONS

- If the average elapsed time from a patient being ready for transport and their arrival at the pickup location is long, it might be because Sunnyvale General doesn't employ the right level and volume of transport staff (a relatively inexpensive human resource). In some cases this situation forces nurses (an expensive resource) to spend valuable time transporting the patient out.

- If, after the patient is transported, it takes too long to turn a bed over, it may be because the hospital didn't employ enough environmental services/housekeeping staff.
- In both of these cases, it's a dime holding up a dollar—a relatively small cost impacting a highly valuable asset: the placement of a patient in the right bed.
- If a patient can't go home because they are waiting to be picked up by family or other transportation, it may be well worth paying for their ride home or creating a discharge lounge where patients can be temporarily hosted while they wait for their ride.
- The hours of operation for certain procedural areas, labs, and imaging functions can impact LOS. For example, a patient admitted late on Friday afternoon may have a longer hospital stay due to test results being unavailable because the lab is closed on weekends. Perhaps keeping the lab open for a half day on Saturday, for example, could help speed up the right placement and reduce LOS.

8. USE PREDICTIVE DISCHARGE PLANNING TO FOCUS CASE TEAMS AND SOCIAL SERVICES

At the end of a patient's stay, many avoidable delays occur around discharge, such as issues with insurance documentation, transportation, and available space at a skilled nursing facility. Many of these delays could be avoided if case managers were alerted to the problem earlier. Historical data regarding avoidable discharge delays can be collected, and a machine-learning framework can be used to identify key case attributes that indicate possible discharge issues. This risk model can then be used as an early warning to alert case managers to at-risk patients. Known factors that often contribute to delays in discharge can be addressed early in the patient's stay, thereby greatly reducing the risk. This will require a shift in the traditional mindset of "peanut-butter-spreading" case-management capacity across all patients in the hospital on any given day.

9. PRIORITIZE SOON-TO-BE-DISCHARGED PATIENTS IN QUEUES FOR LABS/CLINICAL PROCEDURES

Hospitals typically prioritize clinically urgent cases in their lab queues; less urgent cases are usually first come, first served. It might be a good idea to also identify patients who are nearing discharge and place them in a prioritized queue. This can help ensure that lab-test results do not unnecessarily delay the discharge process, thereby blocking inpatient placements. Since most hospitals already have a prioritization process in place for their lab queues, updating the system to include patients close to discharge should be relatively easy.

10. TRANSITION SOME PROCEDURES TO OUTPATIENT

In some cases, it is more cost effective to schedule procedures as outpatient rather than inpatient, particularly if the patient resides locally. It may be helpful to categorically identify certain types of procedures and patients that are good candidates for treatment in an outpatient setting. In some cases where the patient does not live locally, it might make economic sense to move the patient to a local hotel overnight, with instructions to report back in the morning for an outpatient procedure. Of course, this type of decision, due to its clinical nature, requires approval from a medical expert, and it is unclear that an analytic system alone can identify candidate patients/ procedures with the confidence level required. But the system can at least help identify possibilities.

***

Each of these ideas is worth evaluating based on the specific circumstances of any given health system. While some of them will require changing familiar processes and habits, as well as learning and implementing new tools, the juice is definitely worth the squeeze.

Now let's look at a third critical asset in hospitals, one that brings with it a high degree of unpredictability and uncontrollability: emergency departments.

CHAPTER 6

EMERGENCY DEPARTMENTS—DEBOTTLENECKING THE ED

The emergency department (ED) is considered the front door of the hospital. Over half of all hospital admissions,[24] and over three-quarters of unscheduled admissions,[25] come through the ED. Unfortunately, the mere mention of the term ED (or ER) often conjures visions of suffering patients waiting for hours in waiting rooms and examination rooms.

The average patient in the US waits more than ninety minutes to be taken to a bed in the ED and over two hours before being

24 Brian J. Moore, Carol Stocks, and Pamela L. Owens, "Trends in Emergency Department Visits 2006–2014," Healthcare Cost and Utilization Project 227, September 2017, https://www.hcup-us.ahrq.gov/reports/statbriefs/sb227-Emergency-Department-Visit-Trends.pdf.

25 Ashley Gold, "Most Unscheduled Hospital Admissions Come through ER," FierceHealthcare.com, June 25, 2013, https://www.fiercehealthcare.com/healthcare/most-unscheduled-hospital-admissions-come-through-er.

discharged.[26] Patients with broken bones wait a painful fifty-four minutes, on average, before receiving any pain medication.[27] And the number of patients who leave EDs without being seen has almost doubled in recent years.[28]

Long wait times are risky to patient health and detrimental to patient satisfaction, as we have noted before. Sluggish throughput is particularly problematic in the ED, where medical issues tend to be acute. Organizations such as the Center for Medicare & Medicaid Services and The Joint Commission have taken note and have endorsed target metrics around key factors such as wait time, door-to-doc time, and length of stay.

This push from watchdog organizations has provided some impetus for EDs to address these long-standing problems. Process-improvement teams, consultants, and software companies have tried to solve the issues from a number of angles:

- Predicting patient arrival patterns
- Adding beds, machines, and other resources
- "Soft-diverting" patients from the ED via apps, incentives, or telemedicine
- Rethinking triage and vertical treatment methodologies
- Identifying high-risk patients who leave the ED without being seen

---

26 Laura Dyrda, "25 facts and statistics on emergency departments in the US," BeckersHospitalReview.com, October 7th, 2016, https://www.beckershospitalreview.com/hospital-management-administration/25-facts-and-statistics-on-emergency-departments-in-the-us.html.

27 Ibid.

28 "Timely and Effective Care – Hospital," Data.Medicare.Gov, accessed July 2020, https://data.medicare.gov/Hospital-Compare/Timely-and-Effective-Care-Hospital/yv7e-xc69.

- Alerting care teams about ED delays via the use of push software
- Providing dashboards of historical analytics
- Predicting inpatient bed capacity

Very few of these approaches have led to noteworthy, universally applicable solutions. That is because most of these attempts either 1) fail to address the true root cause of the problem, 2) address the root cause but miss an essential piece of the solution, or 3) solve the problem in a way that is unique to a particular ED but is not easily transferable or scalable.

This chapter will provide insights into how to address the ED problem more effectively and universally. But first, as in previous chapters, let's explore the supply-demand problem that lies at the heart of the dilemma.

MATCHING SUPPLY AND DEMAND IN EDS

As in so many other medical service areas, the problem in managing patient flow in EDs comes down to a misalignment of supply and demand. In the case of the ED, there is a disparity between the relatively low supply of ED beds and the ever-increasing demand for these beds.

ED DEMAND

While sources differ as to the exact volume of patients arriving in EDs daily, the trend of increasing demand is inarguable. From 2000 to 2016, the volume at EDs steadily increased from approximately 103 million visits to 143 million visits (CDC, NEDI-USA, AHA).[29] This increase vastly outpaced population growth over the same period: 14.8 percent US population growth[30] versus about 39 percent estimated ED volume growth. Part of the reason for this increase is that, for many patients, the ED has come to represent the most accessible portal into the healthcare system.

The demand problem is unique in the ED for several reasons.

Unpredictable, Uncontrollable Patient Arrivals

Emergencies are by definition unplanned. Whereas most hospital departments and outpatient clinics have the luxury of dealing with scheduled appointments, the ED almost never does. The ED has virtually no control over its arrivals. This doesn't mean that there are no *patterns* to patient arrivals but that the variability is much higher than in other departments. Consequently, the value to be gained from predicting arrivals is harder to realize.

The inpatient bed unit, for example, also deals with unscheduled arrivals. But most of these patients arrive through other hospital departments—namely, the ED and the OR—so the data needed to anticipate upcoming demand exists within the hospital EHR system. The same cannot be said of the ED.

---

29 "Analysis of American Hospital Association Annual Survey Data, 2016," US Census Bureau: National and State Population Estimates, July 1, 2016, https://www.census.gov/programs-surveys/popest/data/data-sets.2016.html.

30 "US Population 1950–2020," MacroTrends.net, accessed July 2020, https://www.macrotrends.net/countries/USA/united-states/population.

A True Twenty-Four-Seven Department

The ED is a 24/7/365 operation. Demand may go down during off-peak hours, but it does not stop. Whereas other departments can opt to end their day late in order to catch up, the ED must rely on a slowdown in patient arrivals in order to do so. This slowdown does not always occur. If an incident causing numerous injuries occurs in the middle of the night, for example, the ED gets further behind rather than catching up. As a result, the next day's operations suffer.

Unstable Patients

Because many patients who arrive at the ED are already deteriorating health-wise, or are at risk of deteriorating, any delay in securing an ED bed becomes not only a patient *satisfaction* issue but also a patient *health* issue. In other words, the demand for an ED bed has a clinical component that supersedes efficiency concerns. A stroke patient cannot wait for a CT scan for the sake of operational efficiency. This reality limits the amount of process optimization that can occur in the ED.

Patients Who Are Already Stressed

Almost no one wants to go to the ED. Patients arriving at the ED are feeling anxious at best and terrified at worst. Because of this stress factor, ED patients have an especially low tolerance for waiting and an especially high bar for satisfaction. Overcoming this patient satisfaction challenge is not trivial. And as we have noted elsewhere, patient satisfaction, as captured in surveys such as HCAHPS and Press Ganey, directly affects reimbursement.

Patient stress levels combined with delays in service can also lead to some patients leaving without being seen (LWBS). While the national average is 2.6 percent of ED patients LWBS,[31] the rates can go as high as 20 percent.[32] Even 2 percent is bad, for both the patients and the ED. Patients who leave without being seen do not receive timely treatment and are at risk of decompensating after leaving—or perhaps of spreading an infectious disease. This risk alone is more than enough incentive to try to reduce LWBS rates to zero, not to mention the PR damage a hospital can suffer.

ED SUPPLY

The main supply component in EDs is beds, though many other resources—such as staff and equipment—are vital as well. The supply of ED beds can be maximized in only two ways: by 1) increasing the physical number of ED beds, or 2) improving patient throughput in the existing ED beds.

Increasing the number of ED beds is extremely expensive. Throughput, however, *can* be improved within the current system resources. Simply put, the usage of existing ED beds should be optimized before new beds are added.

ED-bed throughput is a relatively simple metric of patients per hour departing the ED. Whenever patients are queued up in the waiting room, throughput should be maximized (without compromising patient satisfaction and outcomes). There are a handful of reasons this is difficult to accomplish in the ED:

Patient Care Paths Are Tricky to Manage

Part of what makes the ED unique is the constantly evolving environment. Not only are patients' arrival times largely unknowable,

31 Renee Y. Hsia, "A New Look at Leaving Without Being Seen in EDs," *Physician's Weekly*, March 27, 2012, https://www.physiciansweekly.com/a-new-look-at-leaving-without-being-seen-in-eds/.

32 Ibid.

but their medical issues are largely unknowable too. The ED is a mathematically chaotic system in which small changes in factors such as available resources, patient numbers, and case severities can have a large impact.

The inability to predict and plan in the ED leads to an intrinsically reactive environment. Stopping and deliberating is not a luxury the ED staff often enjoys, especially when it comes to operational decisions. Tasks are often interrupted, and in multiple layers—meaning the interrupting task may also be interrupted. Priorities are constantly shifting. What was a good plan a minute ago is often superseded by a new one.

Moving patients efficiently along their correct path is *especially essential* in the ED.

With such constant interruptions and reprioritizations, combined with the urgency to stabilize and calm patients, it is easy to see why the ED staff has difficulty managing patient care paths. However, moving patients efficiently along their correct path is *especially essential* in the ED. Because incoming patients are unstable and undiagnosed, the longer it takes to get a patient into an ED bed, the higher the chance of an undesirable outcome.

Misalignment Between Departments

Efficiency in the ED often depends on departments outside the direct control of the ED. Misalignment between the ED and the inpatient units is particularly problematic. That is because the ED is dependent on inpatient units for almost all admissions. Sources differ as to the percentage of ED patients who end up being admitted to the same hospital, but the average estimate seems to be about 17 percent. A minority of patients, yes, but *any* stoppage in admissions from the ED clogs the system for all patients.

As discussed in the inpatient chapter, there is often misalignment between the time an ED patient needs to be admitted and the time an inpatient bed is available. As a result, the patient cannot be admitted even when their workup has been completed, thereby resulting in "boarding."

Boarding—the practice of temporarily "parking" patients in hallways as they await admission—is an all-too-common occurrence in EDs. Boarding is more than just a frustrating annoyance. There is a substantial risk that the patient's health will deteriorate in these low-priority boarding environments. The goal of the ED, after all, is to stabilize patients, not necessarily to treat them. EDs do not have the specialized staff and equipment needed to perform care beyond this initial step. Not surprisingly, boarding correlates with increased mortality rates.[33]

Another issue is that, depending on a hospital's size and structure, ancillary resources such as imaging, EVS, and transport may be managed by separate departments and shared among outpatients, inpatients, and the ED. These departments may also have misaligned priorities, resulting in outpatients or inpatients having access to resources ahead of ED patients for a variety of reasons. This situation, too, leads to lower throughput in the ED.

Lack of Coordination Between ED Resources

Lack of coordination between resources in the ED itself often results in decision-making based on incomplete information or on *local* efficiency rather than the efficiency of the whole ED. For example, patients are often placed in ED beds simply because the room is available—without consideration for the workload of the staff assigned to that

---

33 Elaine Rabin et al., "Solutions to Emergency Department 'Boarding' and Crowding Are Underused and May Need to Be Legislated," *Health Affairs* 31, no. 8, August 2012, https://www.healthaffairs.org/doi/full/10.1377/hlthaff.2011.0786.

room. As another example, the lab may prioritize testing on a first-in-first-out basis, without considering the effect that policy might have on a patient's overall length of stay. The fact is that some patients are waiting only on their lab results in order to vacate the ED and should be given priority; others can "afford" to wait a bit longer for labs.

***

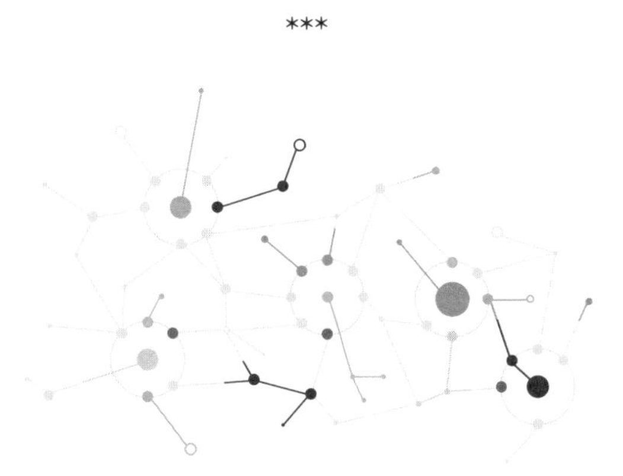

LINKING INDIVIDUAL SERVICES

For the ED to care for patients properly, it must be appropriately staffed to handle the influx of patients at any time on a given day. However, it must also be prepared to address two specific types of linkages. The first is the need for specific tests, such as imaging, bloodwork, and CT scans, which must be completed before a patient can be safely discharged. The second is the availability of downstream resources. In some busy hospitals, it can take upward of thirty phone calls between the ED and inpatient bed units to find a bed for an ED patient. Even after that, it can take hours for that transfer to actually happen. As a result, many ED beds end up being occupied by patients waiting to be moved to an inpatient unit, thus throttling the throughput of the ED.

***

The ED problem can be summarized as a lack of system-level information and coordination, resulting in compromised patient throughput, patient satisfaction, and patient care. Now, let's look at some insights into solving the problem.

INSIGHTS INTO SOLVING THE ED PROBLEM

The solution to the ED problem lies in two main areas: 1) freeing the ED care team from the burden of real-time management of patient throughput so they can focus on patient care, and 2) creating scalable solutions that can be deployed across the whole ED market. The following propositions aim to accomplish both.

1. ACTIVELY MANAGE PATIENTS' CARE PATHS

The goal of active care-path management is to minimize time wasted on a patient-by-patient basis. In many EDs today, the responsibility for managing a patient's care path is dispersed among many team members. This results in the "tragedy of the commons" (i.e., when everyone owns responsibility, no one owns responsibility). The situation is often exacerbated by the misaligned priorities for orders to outside departments, as described above. In summary, managing patient care paths through the ED is often an afterthought, which means longer ED visits and slower throughput.

The Current Reactive Process

The burden of patient throughput usually falls on the ED physicians because often they are the ones judged by the metrics—such as door-to-doc time and LOS—and also because they have the authority to push a patient through. However, the frequent reality is that, after a period of caring for other patients, the physician realizes that certain tasks aren't complete with earlier patients. The physician then proceeds to try to motivate the team to complete those outstanding tasks.

Relying on ED-physician effort to catalyze daily operations is not only cost inefficient but also ineffective. It leads to unnecessary patient wait times and suboptimal decisions.

> Effective patient care-path management requires a system with real-time, ED-wide, analytical oversight—one that actively tracks patients' progress against targeted objectives.

Improving Care-Path Management with Real-Time Operational Insights

Effective patient care-path management requires a system with real-time, ED-wide, analytical oversight—one that actively tracks patients' progress against targeted objectives. The goal of such a system is to keep patients on track to hit key operational metrics, such as LOS and door-to-doc times, without burdening clinical staff. Essentially, you want to create a software-based "pace car" that provides actionable alerts to keep patient care moving along. As a simple example, when all orders have been completed for a patient, the physician might receive an alert. Or if a patient is falling behind in their care metrics, an alert might be sent to the proper party to take action. Active reminders such as this can be quite effective in keeping processes on pace.

The ED director can automatically capture patient information with a simple screenshot and send it to other appropriate departments.

More complex logic can be implemented to alert ED teams as to the ideal sequence in which tasks should be completed. For instance, if the CT has a long queue, the treatment team might be advised to do the consult before the CT. Of course, critical dependencies must be considered when sequencing steps.

Potential Challenges

A major potential drawback to this type of system is notification overload. As important as it is to identify when and to whom to send alerts, it is equally important to identify when *not* to send alerts. Targeting the *proper team member* is crucial to ensuring that the alerts are useful and actionable. Notifying a nurse that the CT is backed up, for example, is not necessarily helpful. Even notifying the CT tech is probably pointless, as the tech probably already knows this. The alert should go only to the team member who can take action—in this case, possibly a CT supervisor who can activate reserve staff or machines.

2. ACTIVELY MANAGE RESOURCE WORKLOAD

Active resource management is crucial too. This means trying to smooth resource workload and optimize resource queues. The goal, once again, is to increase ED throughput.

Staff Workload Balancing

Workload balancing is an attempt to level-load the workloads of human resources tied to ED beds, such as floor physicians and floor nurses. In many EDs today, the main consideration for bed placement comes down to finding an opening that best fits the patient's needs. However, by relying only on this criterion, you risk placing a patient with an extremely busy physician or nurse—thus merely pushing the wasted time from the waiting room to the ED bed. And now the patient is not only waiting but also blocking an ED bed.

A smart system should be able to evaluate nurse and physician workloads in order to suggest the optimal room/bed for placing a patient. Incarnations of this model exist in some larger EDs, where each ED pod is assigned a workload score to balance workloads across pods. This workload score might be based, for example, on a weighted algorithm that considers the number of patients and the Emergency Severity Index (ESI) level. This ESI level then determines which of the pods are busier than others and allocates new ED patients to each pod accordingly.

By bringing higher-level analytics into the picture, the goal is to go one step beyond pods and evaluate the current workload of each individual physician and nurse. The expected workload associated with each *patient* can also be calculated before assigning a bed. Finally, a timing component can be considered. A patient who will be a lot of work initially but who can then be largely left alone may be preferable in a given bed to a patient who has low initial requirements but a high potential for complications.

> **By bringing higher-level analytics into the picture, the goal is to go one step beyond pods and evaluate the current workload of each individual physician and nurse.**

It is also important to try to assess the workload *for each resource* on a patient-by-patient basis. While considering ESI is a good start, it is only a part of what identifies a particular patient as resource-intensive, and it doesn't say much about *when* the patient will need the most attention and *who* will provide it. Scores must be resource-specific because a physician's work and a nurse's work are not equivalent on the same patient.

Other factors, such as the patient's orders and history, can help further define a workload score. It is also important to consider high-attention patients. An autistic patient, for instance, may require a lot of attention from the nurse and/or physician, even with a minor injury.

Optimizing Ancillary Resources

Optimizing ancillary resources—such as imaging, image reading, performing EVS, and transporting—is another important part of resource management. The key to optimizing ancillary resources is to increase throughput on an *ED-wide*, not a patient-by-patient, basis. Taking a page from the manufacturing playbook, high efficiency in individual machines does not directly translate to high productivity for the whole system. Often, to optimize the system, efficiencies at specific machines must drop.

The most important factors to understand about ancillary resources are their queue length, queue order, and throughput. In particular, knowing queue length in tandem with throughput is important. For instance, if a CT scanner is not processing many patients per hour, it may be simply because there isn't high demand. On the other hand, the CT scanner may have a long queue but may also be processing the maximum patients per hour. In this case, there may not be much the CT team can do to speed things up in the short term. However, bottleneck trends over time may make a great case for adding new capacity.

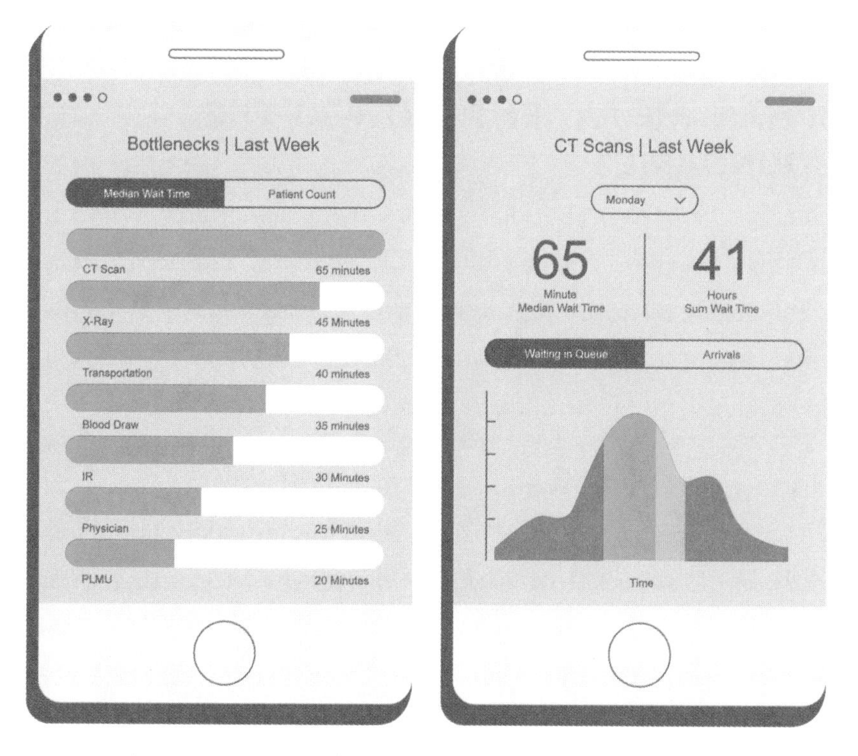

A smart app can look at the analytics for last week, by day and by equipment type, and prompt the ED director to adjust staffing / change schedules / revise policies as indicated.

Being able to *predict*—prior to orders being placed—whether a patient will need to be queued in a particular resource can further enhance optimization. Making this prediction as early as the triage stage can enable each resource to build a forecasted queue, which empowers the resource teams to make earlier decisions about managing their respective resources.

The final piece is queue order—which patient goes when. Again, the goal is to increase *overall* patient throughput. Thus, it might make sense to assign a shorter wait time, for example, to a patient who only has a single order left before discharge than to a patient who may be ready for testing but will be staying overnight. Obviously, all such operational decisions are secondary to clinical decisions. A stroke patient takes precedence at the CT scanner, period.

3. MANAGE PATIENT FLOW AT THE BOUNDARIES

Thus far we have been looking more at managing flow *within* the ED than flow into and out of the ED—arrivals and departures. It is important to realize that although managing *departures* can add value, trying to manage *arrivals* in the ED environment has some downsides that may be difficult to overcome.

Managing ED Arrivals: Inherent Problems

When attempting to manage patient flow, it is important not to focus solely on ED arrivals. To understand why, it helps to separate the issue into two dynamics: prediction and control. ED arrivals, for reasons we've previously discussed, are notoriously difficult to control. Even trying to predict arrivals comes with a set of precautions.

PREDICTING ARRIVALS

An analytics-based system can make some helpful volume predictions. For example, it can use historical data to predict the average number of patients on rainy Tuesday mornings in March. Other factors, such as major events and traffic patterns, can potentially be tracked by the system as well.

The value of such volume predictions, however, is often limited by lack of specificity. While knowing that a wave of patients will be coming in soon may be helpful, this information alone isn't often actionable. Most EDs don't have the option of ramping up staffing and resources on command; the best the team can usually do is brace for impact.

It would be more helpful if patient-specific data could be known—for instance, that approximately twenty patients needing chest x-rays will be coming in by noon. However, this level of insight may be infeasible from an analytics perspective, for a number of reasons, so prediction is not recommended as a primary approach.

CONTROLLING ARRIVALS

The other management option is to *control* arrivals. A few methods for this currently exist. The most common of these is publicly posted wait times. Whether on a billboard or a website, the goal of posted wait times is both to set expectations for potential patients and to possibly divert them to less busy EDs.

Other methods, such as the *scheduling* of ED appointments, seek to actively push low-acuity patients to off-peak hours or to other EDs with more availability. There has not been widespread adoption of such methods, but it is reasonable to believe their impact would be minimal on their own anyway. Why? Because even for low-acuity patients, the primary draw of the ED is speedy treatment and convenience.

The rise of urgent care facilities has been a major initiative aimed at controlling ED arrivals. These walk-in clinics seek to give lower-acuity patients access to timely stand-alone care. The concept can be thought of as patient-elective triage, whereby ED capacity is reserved for needier patients. Urgent care units have been successful in many instances, but the cost and effort of building these facilities is massive when compared to adopting ideas that work within existing resources.

Also, an overarching issue with controlling ED arrivals is that most of these solutions seek to *divert*, rather than *manage*, arrivals. And since hospitals rely on ED patients for revenue, sending patients to competitors' venues is a less-than-ideal solution.

Managing ED Departures and Boarding

Managing ED *departures* may be a more fruitful route to value. As with arrivals, managing ED departures can be separated into prediction and control.

The advantage to predicting departures over predicting arrivals is that the data from the entire patient journey through the ED has already been captured. This data capture makes aggregate predictions and even patient-specific predictions viable.

Before talking about prediction and control of departures, however, it is beneficial to first consider patient *dispositions*—in particular *admit* (send to inpatient) versus *discharge/transfer*. As previously mentioned, admissions make up about 17 percent of ED volume on average but can cause a large portion of the ED throughput problem. Blocked ED beds put enormous pressure on the remaining beds and resources.

PREDICTING DISPOSITION

The major goal of ED departure prediction, then, is to get a handle on admits. Predicting only the *volume* of admits—for instance, there will be twenty admits by noon—isn't especially useful, just as predicting arrival volumes isn't very useful on its own. However, if you can predict the unit the patient will be going to and the time the patient will be moving there, you can create real value. Now specific hospital units can be given volume predictions and even notified as to when to expect patients, so actions can be taken to prepare for any influx.

Ideally, the disposition of each patient should be predicted all along their care path, beginning with triage if possible. As data is collected along the care path, the prediction gains trustworthiness and eventually tips into being accurate enough to be actionable. To take this concept even further, the system might also predict *specific resources* the patient will need when they reach their inpatient bed.

CONTROLLING DISPOSITION

When disposition is not well managed, boarding often occurs. The main cause of boarding is lack of inpatient bed capacity, although other factors, such as lack of transport, can also play in. As discussed earlier, boarding is frustrating for both staff and patients because patients get delayed at the last step in their care path. Another reason boarding is frustrating for ED staff is that while the ED bears the majority of the *risk* associated with boarding patients, the inpatient units have the majority of the *control* to solve the issue—an ED patient cannot be moved to an inpatient bed without the availability of a bed and the consent of the inpatient team.

For that reason, it is important to find ways to not only predict admission volumes and report them to the proper inpatient teams but also to spur the inpatient teams to take action. Coordinating interdepartmental efforts to reduce boarding is vital to aligning priorities across the system and attaining the full benefit of admit predictions.

4. SHARE CARE-PATH INFORMATION WITH PATIENTS

While ED teams often do a great job communicating *health* information to their patients, they frequently overlook *operations* information. For a patient, operational information simply means insight into their own care path. Who is my physician? What is my next step? How long should I expect to wait?

As noted earlier, patients are already in a stressed state when entering the ED. Fear of the unknown only adds to this stress, so anything that can be done to provide information to the patient will help to combat stress and improve patient care.

One way that ED staff currently attempt to provide care-path insight is via the patient whiteboard. This whiteboard consists of basic information such as the assigned care team and the order information. Unfortunately, the whiteboard requires manual updating, which is not a priority for ED staffers who have other patients requiring urgent care.

To make this solution viable, patient information—both operational and treatment-related—can be provided electronically via an in-room device such as a TV/monitor or via a smartphone app. This information can be updated in real time as developments occur. Not only does this solution allow the ED team to focus on patient care rather than information updating, but it also allows patients to be "in the loop" regarding their own care path.

5. TRY TO ADDRESS THE LWBS ISSUE

As noted earlier, one unhappy outcome of combining overstressed patients with overstressed resources is patients leaving the ED without being seen (LWBS).

To help combat LWBS, a prediction model can be put in place

to help identify patients at high risk for leaving. When building such a model, both ongoing and historical data can be mined to look at a wide variety of factors, both general and patient-specific. Are there certain times of day when more LWBS occurs (near dinnertime, for example)? What is the correlation between length of wait and LWBS? Are there certain presenting symptoms that correlate with a greater likelihood of LWBS? Does the patient have a history of LWBS?

While LWBS most often occurs in the waiting room, it can happen at any point in a patient's care path, so the model should track patients through all stages of their treatment in the ED.

If and when a patient is flagged by the system as at risk for LWBS, interventions can occur to help prevent this. Simply talking to a patient and expressing concern and reassurance may be enough in many cases to prevent the patient from leaving.

6. EMBRACE OPERATIONAL INSIGHTS

The final necessary piece to building a value-adding analytics system is to provide a predictive dashboard for long-term insights and planning. This dashboard takes many of the insights mentioned above, digests them in meaningful ways, and presents the information to operations teams in an easily navigable user interface. This dashboard is less for gaining immediate insights—other systems will be in place for that purpose—but more for long-term issue diagnosis and decision-making. For instance, if the ED is backed up every Monday after lunch, how can we address this? The predictive dashboard might identify that the CT is the bottleneck on many Mondays. Drilling more deeply, the user might learn that there is a shift change in the lab around 6:00 p.m. that is causing a knock-on effect on CT queues. This insight can provide the key to a specific, actionable initiative.

The predictive piece also allows for planning to take place *before* such issues arise. Let's say, for example, that for the past three years, there has been a trend toward an increasing number of cranial CT scans in the summer. Armed with this information, the ED team can ramp up its staffing accordingly, or maybe even push for the purchase of a dedicated CT machine before summertime.

***

We have examined some ideas for optimizing the three major assets that form much of the lifeblood and revenue base of hospitals—ORs, inpatient beds, and EDs. Now let's look at a couple of examples of *outpatient* services that present their own unique optimization challenges.

CHAPTER 7

CLINICS—MORE PATIENTS, LESS WAIT

Anyone who has ever visited a clinic knows this simple fact: nearly half of the total time you will spend in the clinic—and often much *more* than half—will be time spent waiting alone.

This clinic "alone time" begins in the waiting room, both before and after check-in. Then there is the time spent perched on the paper-covered bed in the examination room, the time spent waiting for paperwork or prescriptions after seeing the physician, the time spent hanging out in the intake area of the phlebotomy lab waiting to have your blood drawn, and so on. It is not at all unusual to spend an hour or two in a clinic in order to receive less than ten minutes of "face time" with your treating professionals.

Although there has been greater awareness of the problem of excessive waiting times in recent years—and although greater efforts are being made to address this problem—patients are still waiting too long, on average, in clinics.

Wait times matter—for more than one reason. First of all, it goes without saying that physicians and clinic staff want their patients to feel respected and well cared for. But there is also a strong correlation between wait times and patient satisfaction ratings. In a 2018 Vitals study, those providers who had the highest (five-star) ratings also had the shortest wait times—around thirteen minutes—while those with the poorest (one-star) ratings had patient wait times averaging over thirty-four minutes.[34] In this Yelp- and Healthgrades-influenced era, star ratings matter. Ratings, either high or low, can affect patients' choice of providers, especially in locations where there is robust competition. About 20 percent of surveyed patients said they would consider switching providers because of excessive wait times.[35]

But there is also the larger issue of actual patient care. Excessive wait times can affect a patient's decision as to whether to remain in the clinic for the appointment or to leave, untreated. In the Vitals survey above, about 30 percent of respondents said they had left clinics before seeing a physician due to long waiting periods.[36] Crowded waiting rooms can also contribute to the spread of infectious diseases.

Thus, it is both good business practice and good patient care to do everything possible to reduce patient waiting times. Again, this means looking beyond the usual shoot-from-the-hip methods of managing patient flow to more mathematically optimized solutions.

---

34 Thomas Jefferson University Online, "How Patient Wait Times Affect Customer Satisfaction," MedCityNews.com, September 9, 2019, https://medcitynews.com/?sponsored_content=how-patient-wait-times-affect-customer-satisfaction&rf=1.

35 Ibid.

36 Sara Heath, "Long Appointment Wait Time a Detriment to High Patient Satisfaction," PatientEngagementHIT.com, March 23, 2018, https://patientengagementhit.com/news/long-appointment-wait-time-a-detriment-to-high-patient-satisfaction.

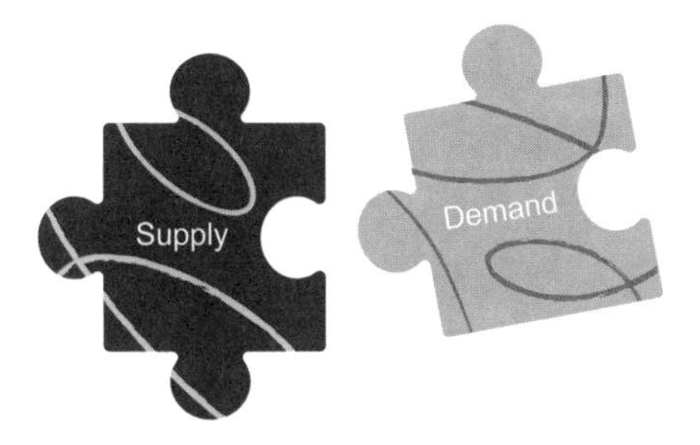

MATCHING SUPPLY AND DEMAND IN CLINICS

The reason waiting times are difficult to control in clinics comes down, once again, to the underlying problem of matching supply and demand. As in many other medical settings, *demand* is highly volatile and unpredictable, while supply is highly constrained and also volatile. Clinics face a particular set of challenges in balancing the two dynamics.

> ***Demand*** **is highly volatile and unpredictable, while supply is highly constrained and also volatile. Clinics face a particular set of challenges in balancing the two dynamics.**

ON THE DEMAND SIDE

The challenge on the demand side in clinics is based on the following factors:

1. **The number of patients who will book appointments with a specific provider on any given day in the future.** This is a difficult number to estimate; there is simply no way to know, in advance, how high the demand for a particular provider's services will be on any chosen day. Depending on the nature of the provider's specialty, demand can also be affected by external factors such as air quality and other environmental issues, localized outbreaks of viral illnesses, allergen levels, weather-related issues (heat exhaustion, snow-shoveling injuries,

etc.), and more. These factors cannot be predicted weeks ahead of time. In oncology, for example, it is impossible to predict, several weeks or months into the future, how many patients will receive a diagnosis for a specific type of cancer and will therefore seek an appointment with the appropriate specialist.

2. **The mix of patient types that will be seen on a given day.** It is also very difficult to estimate how the daily patient mix will break down in terms of new patients, returning patients, and patients coming in for pre- or postprocedural visits. These different types of patient require appointments of differing durations. New patients, for example, usually take considerably longer to see than patients returning for routine follow-ups.

3. **The duration of any particular appointment.** Any given appointment could morph into a more complex medical situation than anticipated, which might cause delays and create a domino effect on the appointments that follow. Because of that and the above issue, it is difficult to accurately predict the duration of each appointment.

4. **Add-ons, cancellations, and no-shows.** Clinics contend with a daily stream of last-minute appointment changes that make it difficult to stick to a predetermined schedule. Absent a sophisticated decision-support system, the frontline staff do the best they can to squeeze in the add-ons or to fill late-cancellation slots with add-on requests. More often than not, last-minute changes only exacerbate an already crowded calendar, leading to longer wait times for everybody.

These factors create a lot of guesswork for employees. Clinical staff struggle to estimate patient load and manage patient flow, based on experience, intuition, and individual logistical skills.

ON THE SUPPLY SIDE

Supply in clinical settings is inherently difficult to optimize for the following reasons:

1. **Each provider is a constrained and unique resource.** Each provider has only a finite number of minutes available in each day, and each provider practices medicine in his or her own way. It is impossible to standardize providers' methods of practicing medicine—nor would it be wise to attempt to do so. Every practitioner in the clinic has perfected his or her craft over many years and has honed a style to match their temperament, skills, beliefs, education, and habits. Some providers talk a lot; others don't. Some write notes as they go; others wait for breaks and periodically catch up on their notes. Some batch all their callbacks at the end of the day; others set aside a few minutes each hour to clear their call backlog as they go. Because each provider has a different set of "dance steps," it is difficult for staff in a multiprovider clinic to devise a standard approach to managing patient flow.

2. **The number and skills of support staff for each provider are unique.** Each provider has a uniquely constructed support team. Some support-staff members are shared among a group of providers; others may be dedicated to a single provider. Some patient appointments require staff with specialized skills; others do not. Efficiently managing the flow of support staff needed for each appointment across an entire clinic is a complex task.

3. **Examination rooms are in fixed supply.** The number of examination rooms allocated to each provider is usually based on some measurement of their "average level of busyness." Therefore, on some days, too few rooms may be allotted to a given provider and on other days too many. The staff end up scrambling in real time to reallocate examination rooms based on actual patient flow—if such reallocation is even feasible.
4. **Important resources are often shared.** Many of the specialized resources relied on by providers, such as scopes, imaging machines, procedure rooms, and specialized support-staff members, are shared among multiple providers in the same practice and may be tied up when a given provider needs them.

Health systems are in a challenging position. They are expected to provide excellent clinical care under all circumstances. At the same time, they have a mandate to be viable from a fiscal perspective. Therefore, they cannot afford to maintain a larger supply of staff, rooms, and equipment than can be effectively utilized on an average moderately busy day. Too *much* supply means their costly assets go underutilized, which results in poor financial performance. Too *little* supply leads to overcrowded waiting rooms and chronically overworked staff, which affects clinical outcomes and brings negative attention and unwelcome pressure to their organization. Striking and maintaining the right balance between supply and demand on a daily basis is therefore essential.

LINKING INDIVIDUAL SERVICES

Many outpatient clinics don't really have much of a linkage problem—the patient sees the provider, receives a prescription, and leaves the health system for the day. However, some specialty clinics (e.g., oncology, neurology, GI) have significant dependencies in both their upstream and downstream nodes.

The upstream nodes might include a lab that collects blood samples prior to the visit or an imaging center that furnishes an x-ray/MRI/CT scan the provider needs in order to make a correct diagnosis. In many cases, the upstream appointment can be completed on a prior day at a satellite facility, which makes the linkage problem easier to handle.

Often, however, there is a downstream appointment that must take place on the same visit—for example, an infusion treatment or a diagnostic procedure such as obtaining a sample for a biopsy. In these instances, the current approach often breaks down in two ways. First, the clinic does not have visibility into the schedule of the downstream node, and the patient is forced to schedule the appointment separately. Second, the clinic simply asserts its preference and sends the patient to the downstream node, assuming the staff will "figure it out" and provide the necessary service. This can result in excessive wait times for the patient, if the downstream node is not in a position to accept an extra appointment at that time.

Therefore, it is crucial to understand the "flight paths" of the segments of patients from each type of provider in each type of clinic who will require each type of downstream service—for each hour of each day (since clinic schedules are driven by specific providers who work on certain days of the week). Downstream services are often more fungible—they usually do not require a specific staff member to provide the necessary service. Understanding these patterns and continually learning from history and tweaking the ongoing schedules is the only way for the downstream nodes to build intelligent templates that can reserve the right units of capacity to enable a smooth flow of patients through the entire health system.

It is crucial to understand the "flight paths" of the segments of patients from each type of provider in each type of clinic who will require each type of downstream service—for each hour of each day.

ATTEMPTED SOLUTIONS OF THE PAST

Administrators and clinic managers have wrestled with the ever-shifting demand-supply problem for decades and have tried countless solutions, including the following:

- Standardizing templates for providers—for example, "Each primary care physician shall conduct four new-patient appointments, ten return-patient visits, and four procedure follow-ups per day." This doesn't work well, for reasons we've discussed. The daily patient mix can vary wildly, and each provider has their own unique "dance steps" that must be followed.

- Hiring additional skilled staff, such as medical assistants, nurses, and physician extenders, to help the provider see more patients each day and to keep the flow of patients moving smoothly. This hasn't worked well either—due to the shortage of skilled staff and the budgetary pressures faced by many health systems.
- Allotting additional examination rooms to the busiest providers, with the idea that a nurse or other staff member will "get the treatment started" and the provider will see the patient as soon as possible. In practice, this has created a "rack and stack" reality wherein patients often wait alone in their examination rooms for long periods of time before the provider finally gets to them.
- Daily huddles in which staff members get together to review the day's emerging schedule, with its add-ons and cancellations, against the day's original schedule, and make on-the-fly adjustments to the best of their ability.

All of the above measures have helped ease wait times to a certain degree, depending on the specific setting and personnel, but remain fairly crude and unsophisticated approaches to a complex, multidimensional problem.

A mathematically based software solution is the only sustainable approach for balancing demand and supply in an optimal way in medical clinics.

STEPS TO A BETTER SOLUTION

A mathematically based software solution is the only sustainable approach for balancing demand and supply in an optimal way in medical clinics. Based on our experience and research, here are the key elements for creating such a solution.

1. DEVELOP A "FINGERPRINT" FOR EACH PROVIDER

One of the most profound insights we struck upon when studying the processes of clinics was this: providers are dramatically different from one another in the manner in which they treat each type of patient. However, they are remarkably consistent *with themselves* as to how they execute each type of patient visit.

The choreography for each type of patient visit is an elaborate "dance sequence" that is unique to each provider. Here is an example:

1. The medical assistant leads the patient to the examination room.
2. The nurse goes into the room and takes the vitals.
3. The resident interviews the patient and writes down his/her observations.
4. The resident reviews his/her notes with the provider.
5. The provider goes in and completes the examination of the patient.

One of these steps may take three minutes, another may take eight minutes, and so on, but the steps need to be performed in a specific order every time. And when a needed staff member is occupied elsewhere, that particular step is delayed.

The unique dance steps of each provider can be compared to the signature moves of a professional athlete, such as a pitcher, a batter, or a tennis player. Once an athlete figures out a windup or a pre-serve routine that works for them, they will always do the same routine in the same way (e.g., bounce the ball three times, lift the head once, and then serve). You can watch a pro like Roger Federer across a five-set match, and he'll do the exact same ritual before every serve. Serena Williams's rituals are entirely different but also self-consistent.

Trying to change an athlete's process to emulate another's would be a fool's errand and would only weaken the player's game. The same is true in medicine. You cannot, and should not, try to change an experienced provider's dance sequence based on efficiency considerations. Imagine, for example, a lung oncologist who studied at Johns Hopkins and has perfected her way of treating patients over a thirty-year period. If you, with no medical expertise, were to walk into her office and say, "I don't think you should be sending the nurse in first. *You* should go in first," you would likely find yourself removing a piece of footwear from your lower anatomy. And deservedly so.

What you're doing is not changing *how* providers see their patients, but rather changing *the order in which the patients show up* to the provider.

So you cannot alter the dance sequence. What you *can do* is capture it, and then model it mathematically so that you can plug any sequence of appointments into this mathematical model and it will tell the provider, "If that's how you schedule your patients, here's how your day is going to unfold; you might want to rearrange your patient schedule." In other words, what you're doing is not changing *how* providers see their patients, but rather changing *the order in which the patients show up* to the provider. This was a major mathematical insight for us early on, and a groundbreaker in developing our approach.

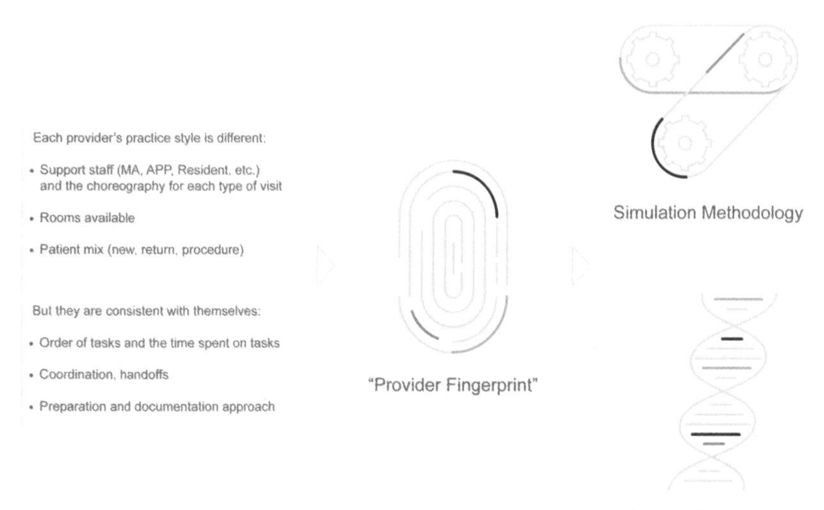

Every provider has a unique "fingerprint" that can be captured and modeled.

The Steps

Here are some steps an optimization team can take toward developing a mathematical model that accurately reflects the choreography practiced by each provider for each type of appointment:

- **Mine the historical data to capture the pattern of patient appointments** on the calendar of the provider. This should include the number of patients arriving at the clinic each day and the type of appointments they have. Understanding the appointment pattern is a first step toward accurately forecasting the likely demand on any specific day in the future.
- **Extract as many time stamps as possible to analyze patient flow** through the clinic. Sometimes this step can be augmented by real-time-location solutions (RTLS) that may be deployed. In other cases, the only time stamps available may be the start and end times of the appointment, as captured in the EHR. Adding a few more time-capture

points to a clinic's process can greatly increase the accuracy of your modeling.

- **Capture the exact sequence of steps** the provider and staff go through to complete each unique type of appointment. Spend some time observing and recording the team's default dance moves and how long each one takes, and then build a basic best-case model, or fingerprint, of how this provider executes each type of appointment.
- **Validate the accuracy of the fingerprint** by running actual historical data through the mathematical model to ensure that the model is able to generate an accurate prediction of how long the average patient visit lasts, what time the clinic concludes its business on a given weekday, and any other metrics you may be tracking.

2. DEPLOY SIMULATION TOOLS

The next step for the optimization team is to build mathematically driven simulation tools that can accurately predict how a day will unfold, given the particular sequence of different types of appointments for each provider in the clinic on that day.

Each clinic deals with a substantial number of last-minute changes that need to be addressed on a daily and a moment-to-moment basis. These include cancellations, requests for new appointments, modifications to existing appointments that may change the duration of those appointments, and so on.

Today, most clinics use a gut-feel approach to solving scheduling problems. They look at a calendar and make intuitive adjustments. As we've said before, "calendar gazing" can work for simple types of appointments, such as those for conference rooms or tennis courts, but it does not work well in settings where each appointment has a variable duration and requires a different set of constrained resources, and in which any decision you make may create a cascading effect on all subsequent appointments.

An accurate simulation model, on the other hand, can enable frontline staff to make intelligent decisions by quickly trying out various scheduling scenarios and observing their likely impact on key operational parameters, such as average visit length and expected end time for the clinic on that day. In a clear and visual way, the user can see, for instance, that if we go with Option A, the staff will be able to go home twenty-one minutes earlier than if we go with Option B.

Not only does the simulation tool provide vastly more accurate predictions than the gut-feel approach, but it also relieves the staff of the stress and burden of trying to perform these calculations mentally, thus preserving their mental bandwidth for more important tasks. A source of interpersonal tension is also removed, as individuals no longer blame one another for making poor gut decisions that adversely affect everyone's workday.

3. MAKE INTELLIGENT POLICY DECISIONS REGARDING SHARED RESOURCES

Most specialty clinics have a number of resources that are, or could be, shared among the providers on any given day. These include physical resources such as examination rooms, imaging rooms, and specialized procedure rooms, as well as human resources such as medical assistants, nurses, and technicians.

In many cases, allocation of these resources is done in the simplest way possible for the administrative team. For instance, each provider may be given a dedicated number of examination rooms based solely on the number of patients they historically see on an average day in the clinic.

A more nuanced understanding of the peaks and valleys in demand-and-supply patterns for each provider, based on sophisticated mathematical modeling, can help clinic staff create an *optimal* allocation of these shared resources across providers. Improving just a few allocation decisions per day in a clinic can significantly streamline the flow of operations.

4. ESTABLISH A "LEARNING LOOP" TO CONTINUALLY IMPROVE OPERATIONAL PERFORMANCE

Each day provides a tremendous opportunity to learn and refine. At the start of the day, the software tool displays a prediction of how events will *likely unfold*, in terms of the expected number of patients to be seen, the types of appointments that will occur, and the predicted utilization of shared resources throughout the day.

At the end of each day, there is new data available on how the day *actually unfolded*. Reviewing this data against the morning's predictions provides a unique opportunity to determine whether any "missed predictions" were one-off events or whether they point to a systemic change in one or more of the underlying parameters. In the latter case, a recalibration of the forecasting, simulation, or optimization algorithm may be warranted. Over time, the math-driven software becomes more accurate through continual tweaking based on real-world data.

UNLOCKING VALUE

The value that can be potentially unlocked by performing the types of steps described above is substantial—although, of course, it varies depending on the unique attributes of each clinic and provider.

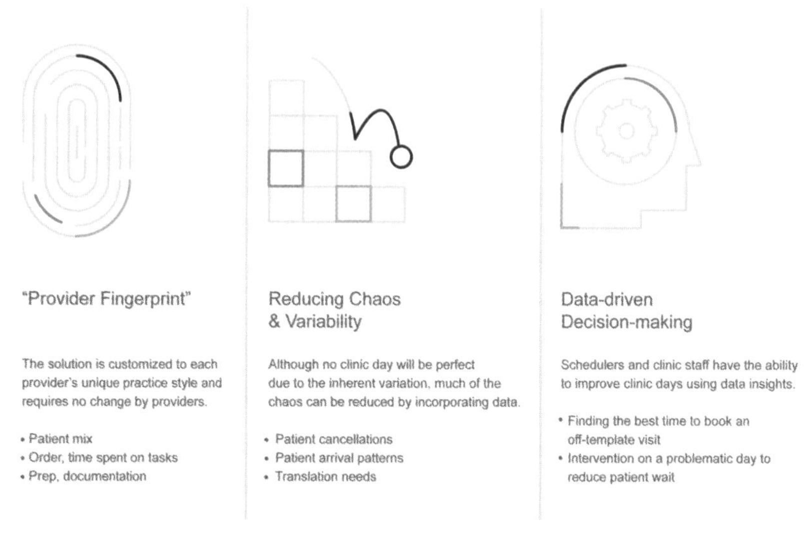

Optimization leads to many benefits—for the patient, the provider, the staff, and the clinic.

Some types of value that can be realized are as follows:

IMPROVED ACCESS TO CARE FOR PATIENTS

Patients' access to healthcare providers is a major issue these days. Patients often spend hours on the phone trying to find an appropriate physician in their area who accepts their insurance plan and is also taking new patients. Many patients feel they have less access to the healthcare system than ever before.

Access simply means the ability to get an appointment with a reasonable lead time. Patients are generally happy when the appointment occurs within a few days to, say, a week or two of the initial contact.

Many providers struggle to free up space on their appointment templates in order to see more new patients. Administrators often set access targets for their providers (e.g., a minimum number of new patients to be seen per week or maximum permissible time period from initial referral to first appointment). Deploying sophisticated methods such as those described above can help clinics and providers optimize their templates, making it possible for them to see more new patients each week without any deterioration in the operational performance of the clinic.

IMPROVED PATIENT EXPERIENCES

Patient satisfaction is directly correlated, as it should be, with the quality of clinical care experienced by the patient. However, one of the biggest sources of *dis*satisfaction—as we noted earlier—is excessive wait times. The issue of prolonged wait times goes beyond one of simple customer convenience. When you, as a provider, see a patient in a timely way, you are honoring the promise you made to them. When a patient can say, "You gave me a 2:30 p.m. appointment, and you saw me at 2:30 p.m.," they feel respected and cared for. They develop trust in you as a caregiver.

An intelligent and dynamic approach to matching demand with supply can reduce the total time patients spend in clinics by up to 25 percent.

FREED-UP EXAM ROOMS

Many clinics find themselves "landlocked," with all their examination rooms preassigned to existing providers for every day of the week. But in order for a clinic to expand, it must recruit *new* providers, and these new providers must be given examination rooms. If the existing examination rooms are being used inefficiently, new rooms must be added. However, this is often impossible or impractical. Any new rooms must typically be placed within—or at least adjacent to—the footprint of the current clinic, since many of the staff and other resources are shared. This forces the clinic to either buy or rent adjacent space, or expand their current building, neither of which may be a viable option at the current time.

It is therefore essential to make the best possible use of the existing set of examination rooms. Optimizing the flow of patients will result in each provider needing fewer examination rooms, thereby freeing up 10 to 15 percent of room capacity to support the recruitment of new providers.

***

Today's clinics face enormous challenges in trying to run efficiently and profitably. They need the most advanced tools available to help them.

One particular *type* of clinic—the infusion center—presents its own special optimization problems, due largely to the length and nature of its appointments. We'll look at that setting next.

CHAPTER 8

INFUSION CENTERS—WINNING THE GAME OF TETRIS

Infusion centers are clinics that provide intravenous medications on an outpatient basis—often for cancer patients undergoing chemotherapy treatment but also for several types of non-oncology patients.

As of this writing, we have spoken with the leadership teams of over five hundred infusion centers over a five-year period, and we have observed that centers all over the country share similar operational concerns. They often find themselves "playing Tetris" when trying to cobble together daily schedules for their infusion chairs in a way that is practical for all parties concerned.

THE THREE MAIN CHALLENGES

There are three challenges faced by virtually every infusion center.

1. Patients tend to **wait a long time** for their infusion treatments—especially in the middle of the day.

2. Chair utilization starts out low and ends low, but has a **midday peak** during which chair *demand* is at or above maximum chair *capacity* virtually every weekday.
3. Infusion-center **nurses miss their lunch breaks** several times each week, and often have high levels of overtime and emergency callbacks on their days off.

On the surface, the scheduling of infusion treatments would appear to be a fairly simple problem, perhaps simpler than other asset-utilization problems in healthcare. After all, the parameters are fairly fixed and predictable. Patients are scheduled in advance. The appropriate mix of medications for the specific treatment is prepared by the pharmacy. The patient is seated in an infusion chair and connected to the infusion pump to start the flow of the medications. The patient then remains seated for the preset duration of their treatment length.

After the initial connection to the pump, all that should be required from the treatment team is periodic monitoring by the infusion nurse to make sure everything is proceeding normally, and the patient is not experiencing any adverse reactions to the medication.

It would seem that scheduling in infusion centers should be as simple as reserving a chair for the patient for the duration of their treatment, much as one might reserve a tanning bed or make a spa appointment. Then why doesn't it work out that way in practice?

We come back once again to our old friends, supply and demand.

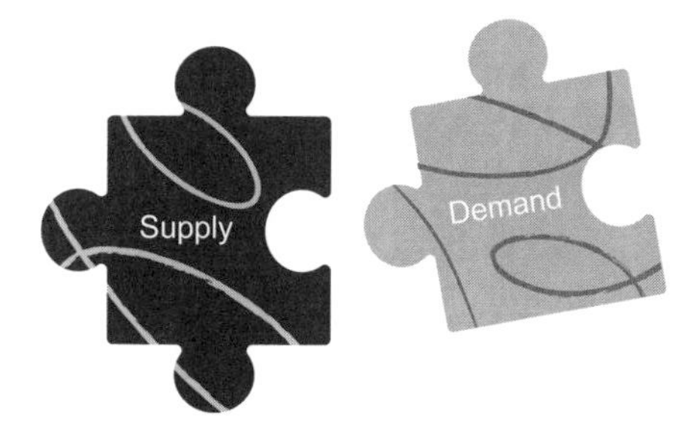

MATCHING SUPPLY AND DEMAND IN INFUSION CENTERS

The fundamental problem of demand-supply matching in infusion treatments comes down to several issues on both the demand and the supply side.

ON THE DEMAND SIDE

The demand side is highly volatile based on the following parameters:

1. **The unpredictable volume of patients on any specific day in the future.** Infusion-center staff don't know how many patients will need to schedule appointments on any given future day. The reason this is a particular problem in infusion centers is that infusion treatments—such as those for chemotherapy patients—are done for each patient in cycles of a specific number of weeks. Receiving infusion treatments on time is not a discretionary thing for patients—so if an unusually high number of patients are scheduled for a given day, the infusion-center staff can't simply say, "We'll schedule some of them for the following week instead."

2. **The impact of cancellations, add-ons, and no-shows.** Patients often need to have lab tests run on the day of their infusion treatment to ensure they are healthy enough

to receive the treatment. Sometimes they do not pass the test, and their infusion treatment must be canceled at the last minute. Other times the insurance authorization is not approved or the clinic fails to confirm the treatment regimen. These situations also result in last-minute cancellations. On the flip side, sometimes a patient needs to receive an infusion treatment right away, as a life-or-death matter. The infusion center has very little control over these situations, and the staff must simply try to adapt the schedule as best they can.

3. **The mix of treatment durations for a given day.** Patients require their infusion chairs for varying durations of time, based on their specific treatment regimens. This is the central issue of infusion-center scheduling that creates a Tetris-like challenge for schedulers. We spoke about this earlier in the book. The number of ways in which a daily schedule can be built for a thirty-chair infusion center that completes sixty to seventy treatments per day of four different duration lengths (e.g., one hour, two to three hours, four to five hours, and six-plus hours) is an integer followed by more than a hundred zeros. As the graphic below illustrates, trying to fit together these "puzzle pieces" of varying sizes in an efficient way creates a logistical challenge that is beyond the capacity of a normal human mind to solve.

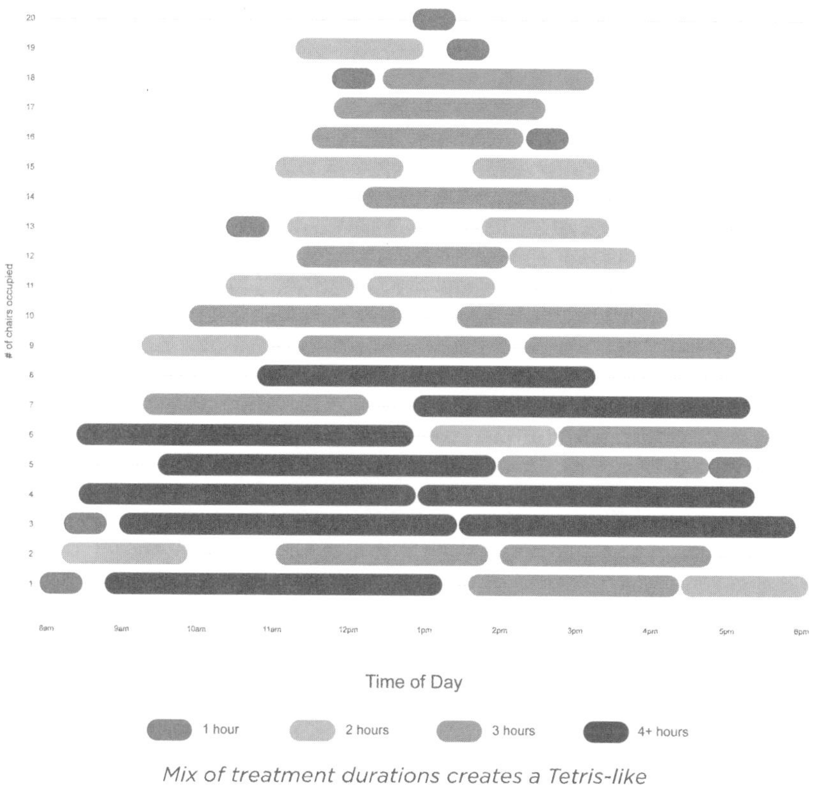

Mix of treatment durations creates a Tetris-like challenge for infusion-chair schedulers.

4. **The likelihood of treatments running longer or shorter than planned.** A variety of factors, such as incomplete chemotherapy orders, the need to add hydration for a particular patient, or problems with lab results, can cause appointments to start later than planned or to run for longer than expected.

5. **The likelihood of patients arriving early or late.** Each late start has a domino effect throughout the day, affecting subsequent appointments. As a rule, lost time cannot be "made up for" later, because the duration of each treatment is fixed and can't be sped up.

ON THE SUPPLY SIDE

The supply side is highly *constrained* (and volatile) based on the following parameters:

1. **The number of chairs and the hours of operation of the center.** Resources are fixed and finite.
2. **Staffing issues.** The number of nurses, the start/end time of their shifts, and the coverage plan for their lunch and other breaks are major factors that affect the availability of human resources on a given day.
3. **Pharmacy issues.** The number of available pharmacy hoods, the hours of pharmacy operation, and the staffing patterns (and staffing irregularities) of the pharmacy affect the turnaround time of drug preparation in a meaningful way. In certain clinics, pharmacies are not dedicated to oncology and therefore have commitments to other departments. In many instances, the pharmacies are not located within or near the infusion center, which can also create logistical challenges in getting the medications to the patient in a timely manner. Oncology drugs are typically formulated uniquely for each patient, are very expensive, and have a very short shelf life. Hence, you cannot premix the medications without incurring a substantial risk that thousands of dollars' worth of drugs will go to waste on any given day.
4. **Policies and practices around patient care.** Infusion clinic policies may determine the number of patients who can be safely managed by each nurse simultaneously and the number of new starts per hour that a nurse may be permitted to handle.

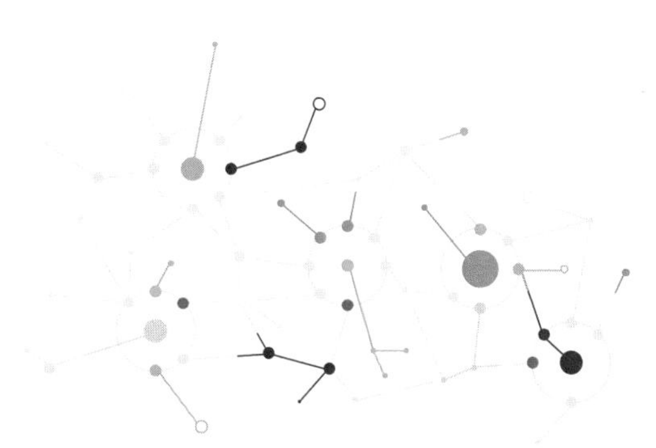

LINKING INDIVIDUAL SERVICES

Infusion treatments—especially for oncology—are complicated by the fact that the patient's ability to withstand the treatment must often be evaluated at the time treatment is to be provided. Therefore, a relatively large number (30 to 50 percent) of patients entering an infusion center need to have been seen the same day by their oncologist.

This creates the classic linkage problem—the patient must first go to the clinic and visit their provider before proceeding to the infusion center. However, clinics often run late and have to deal with last-minute add-ons, cancellations, and no-shows. It is therefore difficult to predict the actual arrival time of the patient. As a result, the infusion center struggles to stay on track with their carefully planned schedule, resulting in long patient wait times and overtime hours for their nurses.

Providers often become frustrated when their patients are forced to endure long wait times for their infusion treatments. They sometimes attempt to "mandate" that the patient receive their treatment immediately following their clinic visit. As noted earlier in the book, this doesn't work either—multiple providers mandating the same time slot will result in the infusion center being unable to fulfill most of the mandates.

Some cancer centers seek to dodge the problem by "unlinking" their infusion treatments—they require that the patient see their provider on the day *prior* to their infusion treatment. However, this is an unsatisfactory solution. It essentially transfers the burden of solving a complicated scheduling problem onto the patient. Imagine if the airlines decided it was too difficult to get bags and passengers on the same sequence of connecting flights. They, too, could dodge the problem by requiring passengers to check their bags a day early and pick them up a day after their flight arrives. While this would undoubtedly make things easier on the airlines, it would impose an unacceptable inconvenience on the flying public. The airlines have therefore taken the needed steps to solve the problem—and health-care systems ought to make the same effort.

Fortunately, the math of large numbers can work in their favor. Infusion centers typically have a large number of chairs and accommodate many patients from many different providers on any given day. Intelligent templates can be designed to optimize the scheduling of the chairs (see example on page 157). Software solutions can also help root out upstream problems. If it is determined, for example, that a particular provider or clinic is negatively impacting wait times for their patients, it is possible to "double-click" on those specific cases. It may turn out that this provider has a habit of clustering all their routine infusion-patient visits into a small block of time (say 8:00 to 10:00 a.m.), thereby creating a large influx of patients into the infusion center during that two-hour period. Each of these patients, who may spend only five to ten minutes with the provider, may require an infusion treatment that lasts several hours. This is the equivalent of connecting a firehose to a garden hose—the flow rate simply doesn't match. Persuading that provider to spread their routine follow-up appointments throughout the day can have a huge positive effect on the flow of appointments in the infusion center and on the satisfaction scores of those patients who were previously caught up in this mathematical linkage mismatch.

***

In practice, most health systems simply ignore the mathematical complexity of matching a highly volatile demand signal with a highly constrained (and volatile) availability of supply. They assume, unquestioningly, that such complexity "comes with the territory" and is therefore largely unavoidable. Therefore, the scheduler simply looks at the calendar and decides that Jane Doe should get the 9:00 a.m. slot for her infusion treatment on a specific day in the future. Or the scheduler agrees to a demand that an oncologist's patient should be seated in the infusion chair immediately after his or her clinic visit.

As a result, virtually every infusion center experiences the midday peak and predictably long waiting times in the middle of the day.

There is a better way of managing infusion-center resources, one that can lead to substantially reduced patient waiting times, improved patient access, smoother ramp-ups and ramp-downs, and a flattening of the midday peak.

There is a better way of managing infusion-center resources, one that can lead to substantially reduced patient waiting times, improved patient access, smoother ramp-ups and ramp-downs, and a flattening of the midday peak.

THE ROUTE TO A SOLUTION

Here are the core elements of a mathematically driven solution that addresses the scheduling and asset-utilization problems of infusion centers.

1. BUILD AN OPTIMAL TEMPLATE TO GUIDE SCHEDULERS

The current practices of "first come, first scheduled" or simply offering up slots based on a calendar do not work. Neither do approaches that attempt to "schedule to a chair" by reserving a chair for each patient based on the time of their appointment or to "schedule to a nurse" by assigning each patient to a specific nurse.

What many infusion-center managers do not realize is that it is possible to build an "optimal template" based on the specific parameters of each facility's assets and operations. This template can be engineered to guide schedulers into intelligently sequencing the start times of the various appointments throughout the day based on their respective durations. Simply put, the template tells the schedulers how many appointments of each duration should ideally be scheduled for each time slot of the day. The objective is to create a smooth ramp-up to attain full utilization of the chairs by midmorning, keep usage flat for the majority of the day, and then smoothly ramp down at the end of the day.

Visually, an optimized template might look something like this:

Optimized Templates

Duration of Treatment (Hours)

Time	0.5	1 - 1.5	2 - 2.5	3 - 4	5+
7:00	2	0	0	0	0
7:15	2	0	0	0	0
7:30	2	0	0	0	0
7:45	2	0	0	0	0
8:00	2	1	0	0	0
8:15	2	1	0	1	0
8:30	2	0	0	0	0
8:45	2	1	0	0	0
9:00	2	0	0	0	0
9:15	2	0	0	0	3
9:30	2	0	0	0	0
9:45	2	1	0	0	1
10:00	2	1	0	0	0
10:15	2	2	0	0	0
10:30	2	1	0	0	0
10:45	2	2	0	0	0

Although the schematic of an optimized template looks simple, there is a lot of math beneath the surface. This is analogous to the "search" box on the Google home page—a simple, rectangular box with a magnifying-glass icon on it. But underlying the simple-looking interface are sophisticated algorithms and billions of dollars of software development work that took decades to build.

Building an optimal template starts with a thorough mathematical combing of the historical data (without patient identifiers) in order to analyze the pattern of patient arrivals by volume, duration of treatment session, and time of day. This enables the creation of a surprisingly accurate forecasting model that predicts the incoming treatment volume and the mix of treatment durations for each day of the week.

Core operational parameters such as nursing shift schedules, pharmacy hours and staffing patterns, number of chairs, policies regarding lunch breaks for nurses, etc., can be translated into mathematical constraints. The optimization algorithm takes these constraints into account and identifies a set of recommended appointment slots for each hour of each day of the week that is most likely to yield a flat chair-occupancy outcome for the middle of the day.

Core operational parameters such as nursing shift schedules, pharmacy hours and staffing patterns, number of chairs, policies regarding lunch breaks for nurses, etc., can be translated into mathematical constraints.

These recommendations from the optimization algorithms are then embedded in the existing appointment templates within the scheduling system to ensure that the schedulers can continue to utilize their current workflow tools.

Of course, real-world scheduling problems will still occur, but at least now you are starting from a best possible position that factors many of these problems into its calculations, instead of from a flawed mosaic that will only become more chaotic as each new problem of the day unfolds.

2. MAKE DATA-DRIVEN DECISIONS FOR TODAY'S REALITY

Regardless of how accurate the optimized templates may be, each day presents a new reality—the volume of patients may be different from the predicted volume, the mix of treatment durations may not perfectly match the expected mix, and/or a scheduler may have been forced to squeeze in an urgent case outside the slots recommended by the template. And of course, a nurse or two may have called in sick.

In order to provide up-to-date guidance to the leadership of the infusion center, a profile of the way today is likely to unfold, given the known issues of the day (such as staff shortages), can be automatically generated at the start of each day. With a precision of ten-minute intervals, this mathematically generated profile can point out times throughout the day when the chair capacity may be tight and/or identify windows of time in which add-on patients would best be scheduled. It also gives the nursing leadership insight into whether or not the day will run smoothly and end on time or whether some level of overtime staffing may be required.

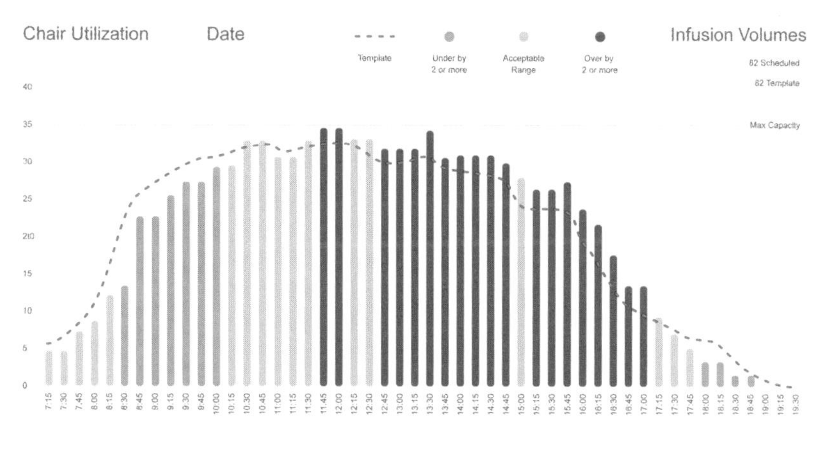

Example of an auto-generated chair utilization profile

Daily nursing assignments ought to be data driven as well. Assigning patients to nurses is yet another crucial decision that is often made based on experience and instincts. This is actually a classic constraint-based optimization problem—infusion nurses are required to be 100 percent dedicated to a single patient at the start of, and near the end of, each infusion treatment. Midflight, however, it is possible for a single nurse to oversee the treatment of a small group of four to six patients, as long as they are in close physical proximity.

As a further complication, some nurses are uniquely qualified to deal with certain types of patients (e.g., bone marrow transplant patients or lymphoma patients) or may be tightly affiliated with certain oncologists/practices. Suddenly the idea of a charge nurse drawing up a spreadsheet a few minutes before the start of each day to make nursing assignments no longer feels like the optimal answer.

What is needed is a constraint-based optimization algorithm. An example of an optimal nursing assignment sheet on a given day might look something like the illustration below.

Optimal nursing assignment example

> *Virginia Oncology Associates is a private practice of forty oncologists across nine offices in the Norfolk/ Virginia Beach area of Virginia. Our largest office in downtown Norfolk sees 250–300 patients per day—*

most of whom receive chemotherapy. We realized we had an additional constraining variable that is an essential part of the way that we provide care. In our model, we assign a given patient to a selected nurse who works closely with the ordering oncologist. This practice has become more beneficial as the physicians continue to become more specialized. We needed to be able to identify the patients being assigned to each nurse to ensure that the subspecialized expertise of the nurse matched the oncologist's subspecialty. It is impossible to depend on a manual approach for getting these nursing assignments right, particularly when many other things need to fall in place, such as the nursing schedule, the volume of cases assigned to each nurse, and the expected arrival time for each patient.

—Nicky J. Dozier, clinical director, Virginia Oncology Associates

3. GROOM THE SCHEDULE IN A PROACTIVE MANNER FOR THE UPCOMING DAYS AND WEEKS

While all the numerous "day of" decisions are important to get right, only so much optimizing can be done at the last minute. In order to consistently run a high-performing infusion center, it is important to be able to look ahead several days or weeks and to "groom the schedule" by making small adjustments that are easy to make *now* but will be harder to implement when the day arrives. These may include rescheduling a patient, suggesting that they receive their treatment at a different location, modifying the shift schedule of a nurse, or placing a restriction on additional appointments being made within a specific time window.

Grooming the schedule requires a sophisticated prediction of how a day several weeks in the future will likely unfold, even though many details of that day are not yet known.

Grooming the schedule requires a sophisticated prediction of how a day several weeks in the future will likely unfold, even though many details of that day are not yet known. In a way, this is analogous, once again, to Google Maps—you can enter a destination and a future date/time, and the app will give you an accurate estimate of the drive time and the optimal route you should take on that future day. Google Maps does not know what drivers will be on the road when you take your trip, which of them will be fast/slow drivers, or which of them will move into the lane in front of you. Yet the app is able to offer a solid prediction based on historical patterns of traffic, weather, and driving conditions on the specified date of the year and time of day.

Similarly, a "huddle calendar" can be generated to convey the information shown in the illustration below. This calendar can provide alert flags on specific days, as well as guidance on the highest-leverage grooming actions that can be taken days or weeks in advance.

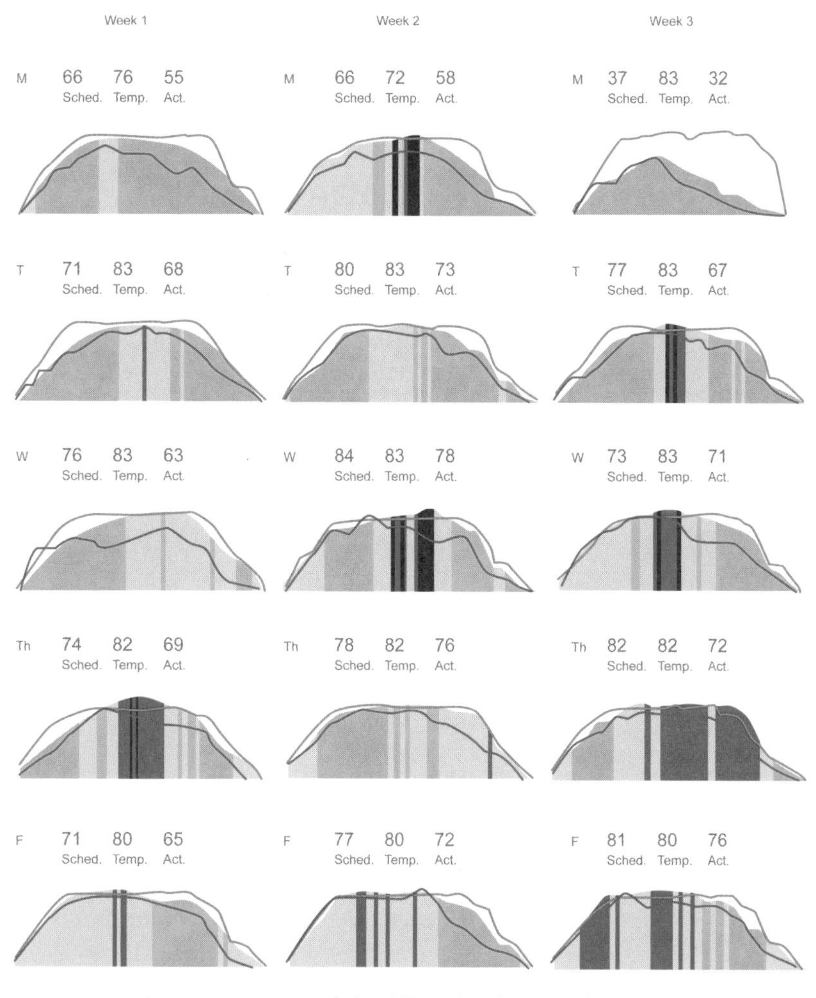

An auto-generated "huddle calendar" can flag problematic days far ahead of time.

4. SOLVE UPSTREAM ISSUES TO THE EXTENT POSSIBLE

In cases where specific providers may be inadvertently creating bottlenecks by sending too many patients to the infusion center in too short a time period, it may be helpful to capture data demonstrating this and share it with the provider. As noted above, this may persuade the provider to consider "declustering" their routine follow-up appointments with infusion patients so that all of those patients don't come in for infusions at the same time of day.

5. BUILD A LEARNING LOOP

Infusion centers are not static; their operational realities change frequently. Providers are added, hours of operation change, new clinical trials are started and ended, nursing complements change, and so on. Consequently, a template that is optimal today may not be optimal six months from now. It is important to keep track of the operational changes that *matter* and to have a systematic, continuous method of mining your data to compare actual performance to predicted performance. In this way, you can identify discrepancies between the two and correctly classify these as either one-off variations or systemically caused issues resulting from a change in the underlying operational facts.

In the case of the latter, you might want to refresh the optimal templates. From our experience, this type of reconciling process ought to happen no more than two or three times per year to minimally disrupt the operational performance of the infusion center.

AN EXAMPLE FROM ONE OF THE LEADING CANCER CENTERS IN THE WORLD

The University of California San Francisco (UCSF) Helen Diller Family Comprehensive Cancer Center is one of forty-seven elite NCI-designated cancer centers in the United States and is one of only two centers in California's Bay Area to receive the prestigious designation of "comprehensive" from the National Cancer Institute.

Across its three campuses in San Francisco, the cancer center conducts over 150,000 patient visits per year, 90,000 infusion visits per year, and 1,500 clinical trials open at any given time. A metric that is routinely monitored as an indicator of patient experience is wait time. A few years ago, the leadership team became concerned that the rapid growth in UCSF's infusion treatments had led to consistently operating over capacity and extended waiting times for patients during peak hours in addition to long lead times to get appointments scheduled.

UCSF had expanded hours, added days to their infusion schedule, and even expanded into additional spaces but struggled to keep up with the double-digit year-over-year growth they were experiencing. They filled the additional capacity much more quickly than they had planned. They were using acuity-based scheduling to help balance the nursing workload but were unable to increase throughput enough to keep up with rising levels of demand. UCSF utilizes Lean methodology for quality improvement, and after analyzing the infusion-center value stream and engaging in a half dozen improvement workshops focused on reducing waste, they achieved incremental changes to overall patient lead times but weren't able to alleviate the significant delays caused by high demand.

In the summer of 2015, Marisa Quinn, director of nursing for Mission Bay and Mount Zion Adult Infusion Centers, the Early Phase Clinical Trials Unit, and the Mission Bay Pediatric Infusion Center at UCSF, and Aubrey Wong, the administrative director for these centers, decided to explore an optimization-software solution. They had exhausted UCSF's internal capabilities and felt that they needed to experiment with something beyond process-improvement efforts and manual manipulation of the infusion schedule.

The two leaders identified a pilot location with fifteen chairs associated with one of their most demanding oncology clinics at the flagship Mission Bay location. They felt that if the effort failed, the damage would be contained. On the other hand, if the effort succeeded at that challenging location, they would have compelling evidence that it could succeed anywhere.

UCSF worked closely with the LeanTaaS software team to optimize the templates for those fifteen chairs. The results were more substantial than Marisa and her colleagues expected—wait times shrank by 30 percent, and the average number of hours operating over capacity was lowered by 40 percent. Suddenly the goal of meeting high levels of patient demand without overwhelming nursing teams and physical resources was much more attainable.

After six months of sustained success, Marisa and Aubrey gained approval to fully deploy the optimization software across all four remaining infusion centers that collectively added eighty-two more chairs. All of the centers went live at the same time, and the results were excellent across the board. At one infusion site, for example, wait times decreased by 10 percent, even though treatment volumes had increased by 43 percent.

In the summer of 2019, UCSF opened the UCSF Bakar Precision Cancer Medicine Building, a new six-story technologically advanced building designed around patients' needs, bringing together researchers, clinicians, and supportive care in one building. Marisa and Aubrey were tasked with consolidating infusion spaces and transferring demand across multiple units in the new building while also adding an additional 26 chairs, bringing the total infusion capacity to 131 chairs across three campuses. Given the success with the mathematically optimized templates, Marisa decided to deploy the software right at the start of operations. As a result, the LeanTaaS software team was pulled in very early on and supported significantly through the planning stages to make the transition seamless for patients and nurses.

***

Infusion centers all over the country, many of which we have worked with directly, are discovering the power of sophisticated mathematics, supported by data-science algorithms and cloud-based software products, to streamline their operations and vastly improve the patient experience.

It was our intention in this section to present a representative sampling of assets and services that epitomize some of the most common optimization challenges in healthcare. There are many other assets in healthcare that can and do benefit from mathematical optimization, and we are constantly expanding our scope into such areas.

Now let's look at how leaders can clear the path toward the digital transformation of their own organizations.

PART THREE

CHAPTER 9

KEYS TO DIGITAL TRANSFORMATION

The US healthcare system, as we said early in the book, is second to none in terms of its innovation, its level of clinical care, and the quality of its people. We have all seen ample evidence of these traits on display during the recent COVID-19 pandemic (we will talk about this more in the epilogue).

However, healthcare organizations face an onslaught of challenges that are affecting their overall efficacy, their public image, their patient satisfaction levels, and their financial viability. We talked about many of these earlier too—an aging population, shrinking reimbursements, human resource shortages, and threats to their core revenue bases, to name a few. Clearly, healthcare systems can no longer afford to allow their operational components to be only "good enough" or to be constrained by the mindsets and habits of the past.

Operational excellence must become the gold standard for healthcare organizations that want to thrive in their current markets and capture new market share going forward. And the way to achieve operational excellence in today's world is through a well-executed digital transformation.

Operational excellence must become the gold standard for healthcare organizations that want to thrive in their current markets and capture new market share going forward. And the way to achieve operational excellence in today's world is through a well-executed digital transformation.

WHY DO ALL OF THIS?

Operational excellence can have a profound impact on the core mission of every health system, which is to deliver outstanding patient care to as many people as possible in a timely and cost-effective manner.

Operational excellence unlocks productivity, the ability to do more with less. An increase in productivity is a net positive for patients, providers, and health systems alike. *Everyone's* time and resources are used more efficiently.

Better productivity means:

Increased access. Moving more patients through the system in the same amount of time makes healthcare more accessible for more people, just as opening more registers at the grocery store allows more people to shop at a given time of day. Seeing more patients within the same hours of operation and/or at the same labor cost is also good for the bottom line.

Lower cost. Greater productivity allows health systems to absorb future growth in patient volume without a corresponding increase in their labor or facility costs. Lower costs lead to greater revenue retention for the company. And when fewer cost increases need to be passed on to patients, the customer benefits as well.

Improving the patient experience. Many of the chief complaints patients voice about healthcare today do not revolve around their clinical experience within the provider's office, but rather around appointment access and long wait times in lobbies and examining rooms. Increased productivity allows health systems to offer more appointment choices today, tomorrow, and this week, and to reduce patients' wait times throughout the course of each encounter with the system. Improved patient satisfaction scores translate to increased reimbursement levels and better patient relationships.

Improving utilization. By increasing the use of scarce, expensive assets, systems can defer the need for additional capital to expand buildings or buy more equipment. Providers and patients are able to access specialized equipment, such as MRI machines, in a timelier manner, which can lead to better clinical outcomes as well as reduced patient anxiety.

Competitive advantage. Improving the patient experience allows the health system to position itself as the "destination of choice," particularly in competitive markets where patients have the choice of several world-class institutions within a one-hour drive. The lifetime economic value of a patient is high—for many patients, their initial clinic visit leads to future visits, procedures, and inpatient stays. Patients, for their part, are looking primarily for excellent clinical care and secondarily for a good customer experience. The ability of a health system to deliver both of these outcomes engenders strong loyalty, which means that the relationship with the patient can be nurtured for years or decades to come. Happy patients recommend family members and friends.

Financial sustainability. Increased efficiency and productivity improve the financial health of the organization in a world where reimbursement levels are falling and operating margins of virtually every health system are under pressure. The financial stakes are high, and even "small" gains in efficiencies can have a substantial payoff. For example, a 2 to 3 percent improvement in the prime-time utilization of operating rooms can be worth $200,000 per OR per year for most systems. For a midsized health system with one hundred operating rooms, this kind of gain represents tens of millions of dollars in annual revenue, which can be redeployed to further strengthen the health system. An inpatient bed is an asset with a revenue potential of $2,000 per bed per night; optimizing the usage of these beds by even a modest amount can have millions of dollars' worth of impact each year.

WHAT HAS HELD HEALTHCARE BACK?

Operational excellence is hardly a new concept for healthcare—we have seen many efforts toward that goal in recent years—but there have been some persistent roadblocks to its full realization. Having worked with over one hundred large health systems representing three-hundred-plus hospitals, and having spoken to leaders at over two hundred others at length, we have identified some commonly held "core beliefs" that may be holding healthcare back from implementing the actions needed to achieve true operational excellence through digital transformation.

> We believe it is time for health systems to understand and learn from these other industries by challenging long-held beliefs that may be holding them back.

As previously pointed out, many other asset-intensive businesses (e.g., transportation, retail, airlines, hospitality, food services) have made dramatic progress toward transforming their core processes over the last decade. Good models of digital transformation now exist for all to see. We believe it is time for health systems to understand and learn from these other industries by challenging long-held beliefs that may be holding them back.

In 1997, Robert Kriegel and David Brandt wrote a book entitled *Sacred Cows Make the Best Burgers*. In it they argue that many businesses harbor sacred cows—outmoded ideas and practices that somehow acquire "untouchable" status and prevent companies not only from *keeping up* with change but also from *getting out ahead* of change in a competitive way. Healthcare executives would benefit greatly from tackling their own sacred cows and launching ambitious initiatives to transform their institutions from an operational perspective. This means looking at the best digital solutions out there and then making the needed internal changes so that these solutions can be implemented as rapidly as possible.

Below we will look at nine core beliefs we believe are currently holding health systems back, and then we'll explain our perspective on the reality of each of these beliefs. The nine core beliefs fall into three broad categories:

- The need for change
- The role of IT, innovation, and process-improvement teams
- The process of selecting and managing a portfolio of initiatives

I. THE NEED FOR CHANGE

Core Belief 1: *"Healthcare is its own animal, and therefore the rules that apply to other businesses don't apply here" or "Our health system is unique (geography, size, range of services, patient population served, etc.) as compared to others."*

Reality: Operationally speaking, healthcare is fundamentally no different from any other asset-intensive business that has substantial demand-and-supply stochasticity. And while health systems may indeed have unique characteristics relative to one another, they share far more in common.

The real issue here, as we see it, is not so much that healthcare is unique as that it has been historically protected, to some degree, from market forces. Here are some examples:

- It is difficult to open a new hospital.
- Patients don't have a great deal of choice in health systems within a catchment area.
- There has been very little price transparency in the industry; thus health systems have been able to "name their own price" for many of their services.

Most other businesses have traditionally been more exposed to market forces and to direct competition, and have thus been *forced* to strive for operational excellence—digitizing, optimizing, streamlining—in order to contain costs and improve customer service. Healthcare, until recently, has had the luxury of writing many of its own rules and allowing some of its sacred cows to stand. It isn't accustomed to viewing the market as a *prime* consideration.

Now, however, for reasons we have looked at already, economic forces (reimbursement pressures, lack of capital, etc.) are bearing down on healthcare. And as with any other industry, when this happens, the invisible hand of the market will eventually have its way. This means healthcare must begin to think more competitively and innovatively. The market differentiators in healthcare will revolve around better patient access, lower cost, and superior quality of service.

Core Belief 2: "*Our EHR ought to be able to accomplish these* *objectives."*

Reality: Yes, you spent a *lot* of money on your EHR, and perhaps you were promised a lot of functionality and problem-solving to go along with it. EHRs have been a vital addition to healthcare operations by being the repository that is the "single source of truth." But your EHR is not going to perform high-level predictive analytics for you. Health systems need analytic "speedboats" to complement their EHR data repository that is like an "aircraft carrier."

In order to optimize operations, you must go beyond dashboards and alerts and begin employing high-value analytical tools that can offer mathematically optimized solutions to specific and complex supply-demand problems. You must use technology that places the right analytics, insights, and recommendations in front of the right users (surgeons, schedulers, nurses, executive teams) at the right time, via simple mobile and web interfaces.

Core Belief 3: *"We have internal models and homegrown tools, and they are good enough."*

Reality: Homegrown tools typically *aren't* good enough. We have seen health systems with five different service lines, each using its own internal versions of Tableau reports and Excel spreadsheets, and each relying on models that measure different things and employ different terminology. These in-house tools are created by well-intentioned people for specific purposes, but they don't usually scale across time, across providers, and across use cases.

Many homegrown tools and models are good as a proof-of-concept exercise—but they don't have wide applicability. In order to scale, a tool must be easily usable by hundreds (or thousands) of people simultaneously, on a continuous and reliable basis. To effectively deploy such tools requires a significant, ongoing investment in building, maintaining, and refining a suite of software products and the underlying technology infrastructure. Such development may be beyond the capacity of even the most sophisticated in-house teams.

II. THE ROLE OF IT, INNOVATION, AND PROCESS-IMPROVEMENT TEAMS

Core Belief 4: "*IT should take the lead on digital transformation.*"

Reality: The healthcare industry tends to rely on IT more than perhaps it should for digital innovation. This is understandable; healthcare professionals want to focus on providing good clinical care, not on software and technology. However, in the rest of the business world, the role of IT is understood to be one of providing infrastructure, security, and policies to implement business transformation tools, not one of solving complex operational problems. IT can be a huge enabler but cannot possibly know the details of every part of the health system's business—or the functional needs of its frontline staff—well enough to take the lead on digital transformation.

A strong relationship between the business unit and the IT unit is critical. The two units must depend on each other to be experts at their roles. If management wants to achieve a certain business goal and decides to spend the money on implementing a system-wide solution, IT must partner with them to help them think through their available choices, but IT ought not be calling the shots. The cart shouldn't be pulling the horse.

Business leaders must *own* the movement toward digital transformation, which may mean fundamentally retooling core processes—such as manually operated, influence-based OR block policies and heuristically developed scheduling templates—that have remained unchanged for decades. All such processes, and more, can be radically improved through analytically based digital transformation.

Every service line, every process owner, every asset scheduler must be committed to using these advanced tools in combination with the data in their existing repositories (such as EHRs) and their own human wisdom to make data-driven decisions. The tools must become a central aspect of operations, not an "extra" feature that only the tech-savvy team members and IT people choose to embrace.

Says Steve Hess, the CIO of UCHealth:

> "We have brilliant physicians, excellent clinical and administrative leaders, and good process-improvement capabilities. However, achieving operational excellence in healthcare takes more than just people. To really move the needle, we needed to deploy advanced algorithms that highlighted specific opportunities, and we needed the ability to combine predictive analytics, machine learning, and optimization with a process-improvement mindset resulting in actionable interventions that made it easy and intuitive to hardwire. We had to connect the dots between the data, the learning, and the sustainable change needed."

Core Belief 5: "*Our innovation team will handle it, and we already invest in start-ups that can solve our problems.*"

Reality: Innovation is the job of everyone, not just a special team. Setting up a dedicated innovation group is a great first step, but organizational change happens only when a significant portion of the health system's leadership and workforce is open to improvement and enthused about innovation. In order for widespread buy-in to occur, the positive, practical aspects of the innovation must be

> **Innovation is the job of everyone, not just a special team.**

seen and felt by real people. Lives and jobs must be made easier by the change, not more complicated.

Funding start-up tech companies can be a great step in the development of new technology. But for start-ups to do well, they need strong healthcare partners that can actively help them fashion products that relieve hard pain points that other health systems are also experiencing. This is the only way the developers' efforts will have true ROI.

If you fund start-ups, make sure you push them to build "painkillers" and products that scale. Eventually these start-ups will need to win the bigger market, not just the handful of health systems that invested in them.

Until we empower and *require* departments and revenue/cost owners to innovate, do more with less, and explore new ways to solve problems in scalable ways, not much will change. Culture is critical in this regard. The culture of your organization must be one in which innovation is sought and honored.

Core Belief 6: "*We have invested in process-improvement methodologies (e.g., Lean, Kaizen, Six-Sigma, etc.) and are therefore well positioned to execute a transformation.*"

Reality: Lean/process-improvement efforts usually place their emphasis on teaching leaders to lead by going to the "gemba," sharpening their observational abilities, gathering facts on the ground, and trying to solve problems at the root-cause level. Lean consultants are trained (and therefore train their mentees) to be suspicious of computer models and problem-solving that occurs while looking at a laptop in an office. This suspiciousness has some validity, but the fact remains that there are certain tasks only an intelligently designed computer program can accomplish.

If your fundamental demand-supply matching methodologies

(e.g., OR block allocation or infusion-chair templates) are suboptimal, no amount of process improvement around them will create much value—there is only so much even a world-class Scrabble player can do if he consistently draws only vowels for tiles. While many aspects of the workflow *can* be improved using process-improvement methods (e.g., increasing the percentage of patients who show up with complete and accurate orders), it is simply not humanly possible to account for the huge variability, volatility, and unpredictability of the patient arrival signal without using sophisticated mathematical models. Imagine Uber relying only on process-improvement techniques to pair millions of daily riders with drivers in a timely manner.

Purely process-based excellence initiatives are best reserved for addressing aspects of the workflow that involve human interactions and the establishment of standard work procedures; they don't really move the needle much when it comes to solving the core mathematical problem of a mismatch between demand-signal patterns and the constrained availability of supply (people, equipment, rooms, etc.).

Supply-and-demand matching must be optimized via mathematically sophisticated means. However, this optimization can (and should) coexist harmoniously with other process-excellence initiatives. Many of the core concepts of process excellence (e.g., Heijunka or level-loading, one-piece flow, Kanban, Takt time, matching production to the demand signal, etc.) can and must be embedded in the optimization algorithms along with the mathematical intelligence.

The effectiveness of coaching frontline leaders in Lean process flow is, in fact, dramatically amplified when the underlying assets *have first been optimized* by mathematical methods. Give the Scrabble players a decent rack of letter tiles to start with, and *then* you can teach them the finer points of how to play a winning game.

III: THE PROCESS OF SELECTING AND MANAGING A PORTFOLIO OF INITIATIVES

Core Belief 7: *"Once we decide which direction to go in, IT can procure the solutions."*

Reality: Most operational leaders have never had to procure modern software tools directly from vendors. Historically, most tools have been purchased via big, enterprise-wide contracts with EHR vendors that have been handled by the IT department. Hence, most leaders don't really know how and when to engage their legal, procurement, and IT teams in order to execute such decisions rapidly and cost-effectively. This mindset may need to change when it comes to finding the right tools to solve complex, context-specific healthcare management problems.

IT is a key service provider to a business. IT can be a huge enabler and facilitator of new technologies, but when it comes to choosing and incorporating predictive analytics tools with very specific healthcare applications, the whole organization must be actively involved, especially the operational leaders.

Core Belief 8: *"Each initiative must be able to demonstrate a clear and tangible ROI that can be attributed to that specific initiative in an unambiguous manner."*

Reality: Most of the methods health systems use for measuring potential ROI on digitally based initiatives are old and based on procuring legacy software systems that have often been deployed on-premises. A much deeper understanding of web-based tools, SaaS (software as a service) business models, and the key drivers of value needs to be in place in order to build an accurate ROI model that can be fine-tuned to the particular business process being looked at.

ROI is not just about reduced costs and increased revenue. There are many other important metrics of success, such as improving patient access, reducing patient wait times, and making it easier for independent physicians to work with the health system.

Some initiatives are "enablers," in that they allow other initiatives to have an even greater impact than they might have had on their own. In such cases, it is often impossible to attribute a specific ROI to the enabling initiative. A golfer may consistently hit the ball 250 yards down the fairway by keeping her head steady, focusing her eyes on the ball, keeping her feet in position, and swinging the club smoothly. It is impossible, after the fact, to specifically attribute thirty of those yards to keeping her head steady. Demanding that each initiative show a discrete ROI can prevent leaders from implementing holistic and intuitive solutions that can work together to improve the system.

Core Belief 9: "*We want to move fast, so let's launch numerous initiatives.*"

In an attempt to move quickly, health systems often launch more operational initiatives than their resource and leadership bandwidth can effectively handle.

Launching a dozen initiatives, each of which is incomplete or partially successful, is usually far less impactful than staging a few well-chosen initiatives at a time and driving them each to a successful outcome. The process should be more vertical than horizontal. Make deeper, more meaningful changes in a few key areas that are likely to have the greatest practical impact in the shortest possible timeframe.

Rethinking "moving fast" in this way requires a willingness to "fail fast"—to determine quickly whether an intervention is working or not. If a promising-looking initiative starts to flounder, it may be time to kill it and move on to the next one. In our dealings with health systems, we have seen, time and again, initiatives that have languished for years without clear evidence that they have delivered any meaningful and lasting change.

THREE PILLARS FOR ACHIEVING OPERATIONAL EXCELLENCE THROUGH DIGITAL TRANSFORMATION

Achieving operational excellence in any industry usually requires transforming the workflows that drive core operational processes (e.g., scheduling, asset allocation, reporting, policy making). The three fundamental pillars of such a transformation are:

- Articulating a clear mandate for change
- Establishing effective governance structures
- Demanding disciplined execution of key initiatives

Let's look at these three pillars as they apply specifically to achieving operational excellence through digital transformation in health systems.

A CLEAR MANDATE

The commitment to operational excellence starts at the top of the organization. The senior leaders of the health system must articulate a compelling vision of the benefits that will be realized through operational excellence. The more this vision is informed by a true understanding of the pain points encountered by the frontline staff on a regular basis, the easier it will be to achieve buy-in from the workforce. "We know that many of you have been working through your lunch hour two or three days a week, and we are committed to solving that problem." It is important that higher organizational goals are also included in the vision.

A digital transformation signals a clean departure from the "old way" of doing things (i.e., making small incremental improvements to existing processes). Hence, it requires a clear and well-articulated mandate, a willingness to commit the necessary resources, a timetable for the transformation, and a road map for achieving the vision.

Senior leadership needs to stress the importance of solving these problems *at scale*—and not in a one-off, Tableau-dashboard sort of manner. Solutions must be robust, universal, practically valuable, easy to navigate, and visually appealing. After using these tools and processes for a brief time, staff should never want to go back to the old way of doing things.

Along those lines, not every solution can or should be developed in house. After all, even the largest corporations, with billions of dollars in their IT budgets, don't waste their resources trying to create and maintain, say, a spreadsheet program—they simply buy Microsoft Excel and provide it to all of their employees.

As a general rule of thumb, health systems should do the following:

- **Build** solutions to problems that their internal resources are uniquely qualified to solve—either because of their expertise, their access to relevant experts, or the privileged/confidential information they possess.
- **Partner** with third-party companies that have complementary technology or solutions that might be able to amplify the impact of the health system's scarce internal resources.
- **Procure** solutions where the size and competitiveness of the external market has created a distinctive solution at a scale that no single health system can hope to replicate.

What you can do:

- Model a mindset of embracing change initiatives that are supported by well-designed digital tools and that show tangible value in improving operational performance.
- Effectively communicate the vision for the initiative *before* the

change, *during* the change, and *after* the change as the bugs are still being ironed out. Consistent, long-term messaging across multiple media (written, verbal, video, etc.) is critical to the success of any change initiative.

EFFECTIVE CROSS-FUNCTIONAL GOVERNANCE

Health systems are complex institutions with many diverse stakeholders. Unlike in most corporations, the senior teams are usually blended from two very distinct tracks—clinical and administrative. In academic medical centers, there are also faculty aspects from the school of medicine to be considered. Hence, the decision-making process is often amorphous and not well understood, even within the organization itself. Often there is no single individual who has the unilateral authority to say "yes" to a particular initiative, but there are many individuals from many different disciplines who can say "no," thereby stymying progress.

In order to achieve the desired vision for operational excellence, leadership from the clinical, administrative, and technology functions must be fully aligned with the vision and prepared to commit the time, energy, and resources needed to make it successful.

In order to achieve the desired vision for operational excellence, leadership from the clinical, administrative, and technology functions must be fully aligned with the vision and prepared to commit the time, energy, and resources needed to make it successful.

Therefore, a clear cross-functional governance mechanism must be established at all levels of the health system—from the senior leaders who form a steering committee that strives to eliminate implementation barriers to the departmental leaders who hold the owners of each specific initiative accountable for the progress their teams are making.

There must also be a commitment to providing the infrastructure needed to enable frontline employees to make data-based decisions. This means investing in small, nimble tools, department by department, not just in massive projects that may take years to see fruition. Focus on practical chunks of innovation to core workflows that have the highest potential leverage—for example, intelligent scheduling templates and automated bed-allocation guidelines.

Operational excellence is not mainly about improving the productivity of the people performing routine workflow tasks, although that can help; it is about making better daily decisions that can improve the utilization of assets that are worth millions of dollars or that affect the flow of thousands of patients each month.

In any large health system, there are hundreds of frontline people, each of whom might make a hundred "microdecisions" each day. These decisions are often based on gut feel, anecdotal evidence, or legacy judgment from having "always done it this way." Augmenting even a small portion of these daily microdecisions with AI, machine learning, and recommendation engines can have a profound impact on the operational performance of the health system. Intelligent recommendations flow from going deep, not wide. Sprinkling a few dashboards and alerts across the whole enterprise will not result in higher-quality decisions. However, drilling deep into each of the many factors that influence any specific decision and identifying the exact data fields that need to be populated, as well as the computational logic that binds these disparate data fields together in meaningful ways, can result in consistently helpful automated recommendations that augment (not replace) the judgment of the relevant frontline person.

Department leaders need to be ruthless about the metrics that matter to the operational performance of the health system. Examples include the following:

- Prime-time OR utilization
- Length of stay
- ED volumes and wait times
- Daily volumes and wait times for a variety of patient encounters such as labs, imaging, infusions, etc.

Leaders need to hold the frontline staff accountable for hitting these metrics and must provide the tools and training for them to accomplish this ongoing goal.

What you can do:

- Consider allowing projects to be established and funded by the relevant business owners. Let the heads of each department/function make informed and intelligent decisions about when to use people and when to use technology. But hold them accountable for hitting the needed metrics.
- Establish stretch goals to stimulate higher aspirations and more creative solutions than the typical "Let's hire more people" or "Let's generate another report."
- Leverage your critical mass of resources against problems that can show ROI in six to twelve months.
- Solutions should have universal meaning and usability across departments. A particular spreadsheet that worked well for a one-time analytic might not have generalized applicability. When designing system-wide solutions, you need to provide a good user experience, the right training, the right workflow

parameters, and a high-availability software service.

- Establish progress-review committees of cross-functional leaders that can manage the sequencing and staging of individual initiatives within the overall portfolio and can review the progress being made on each initiative at any point in time.

DISCIPLINED EXECUTION

The transformation to operational excellence should occur in a tightly focused manner. This might mean partitioning key initiatives into phases of short duration, for example, twelve to sixteen weeks. Grandiose initiatives that drag on for months or years, without a clear definition of success, kill momentum. The killing of momentum is further exacerbated by the changing composition of the team, as people join or leave the organization or take on different roles within the health system.

It is wise to be selective about the resources that are assigned to a specific initiative. Often, health systems will have the same few individuals working on ten different initiatives at the same time. Human beings are simply not built to "context-switch" to such a great extent. This leads to the familiar scenario in which everyone is so busy attending meetings that nothing gets done between the meetings. Eventually, these initiatives devolve into "talking about talking about getting something done," rather than actually doing anything.

Focus accomplishes "miracles." During the recent COVID-19 crisis, one health system executive confided to us that they had been talking about moving specialty clinic appointments to a telehealth platform for the past three years. Then, within a matter of three weeks, 70 percent of their oncology visits were being conducted over a virtual platform. Such is the power of focus.

Once you have established tightly defined phases of twelve to sixteen weeks each for each initiative and assigned a focused team (i.e., contributors working on no more than two to three such initiatives at the same time), you can then begin to define work plans with specific milestones and deliverables. This will enable the cross-functional progress-review committees described above to monitor the progress being made and to offer teams the right combination of challenge and support to keep the initiative moving forward.

It is crucial to deploy tools and technologies that will advance the desired outcomes for your initiatives.

Again, it is crucial to deploy tools and technologies that will advance the desired outcomes for your initiatives. In our personal lives, we adapt rapidly to tools that make us more productive—almost all of us have some combination of Lyft, OpenTable, Amazon, and banking apps on our smartphones. Think of tools as *amplifying* the expertise of the process-improvement teams and subject matter experts, not replacing it. Even master craftsmen can benefit from having laser-guided cutting tools to help them build bigger and better masterpieces at an accelerated pace.

What you can do:

- Make the effort to identify the best tools and technologies for accomplishing your digital transformation—SaaS has made everything much more affordable.

- Augment your human expertise by training key staff to use the tools.

***

Healthcare companies cannot keep *spending* their way out of trouble by investing in more and more infrastructure; instead, they must *optimize* their way out, by making better use of the assets currently in place. Hospitals and clinics today face the same cost and revenue pressures that retail, transportation, and airlines have faced for years. As Southwest, Amazon, FedEx, and UPS have demonstrated, asset-intensive and service-based industries must maximize efficiency and do more with less if they wish to remain viable. Healthcare must do the same. When health systems handle this transformation intelligently and competitively—by harnessing the power of high-level math—their patients and providers benefit as well. A true win-win-win is achieved. And better healthcare is the result.

EPILOGUE

DEALING WITH UNEXPECTED DEMAND-SUPPLY SHOCKS (E.G., COVID-19)

When we started writing this book, the world was in a "peacetime" economy in which the normal rules of supply-demand matching in a steady state could be applied. However, halfway through the writing, we and the rest of humanity were confronted with the most unexpected circumstances our generation has faced thus far—a "wartime" scenario in which the spread of the COVID-19 virus created an enormous supply-demand shock to the healthcare system, a shock we are still working our way through.

Although all the principles and examples we discussed in the body of this book remain valid and will presumably become the norm again, the COVID-19 situation has thrown healthcare into an enormous temporary imbalance. In this epilogue we will look at how, even in such an extreme situation, the powerful use of analytics-based intelligence can help health systems recover faster as they grapple with severe demand-supply "shocks."

WHAT HAPPENS DURING A SHOCK

During a period of economic and social shock, demand for certain assets and services can fall through the floor in the face of fixed, expensive supply shocks. We saw this happen with travel after 9/11—no one wanted to fly for a while—and with mortgages after the 2008 economic crisis. The difference with healthcare is in how the backlog will need to be addressed. When the demand for air travel slowly returned, there was no real backlog to be dealt with. The flights that didn't happen during the slowdown simply went away. And when the housing market came back, the backlog of house purchases was worked through slowly and organically.

The big difference with healthcare shocks like COVID-19 is that the backlog of needed services built up during the crisis needs to be addressed in a short period of time once the main crisis is over. All of the elective surgery that was postponed during the COVID-19 pandemic, for example, still needs to occur after the crisis is over, and in a timely way.

NATURE OF THE COVID-19 DEMAND-SUPPLY SHOCK

As we write this, in the summer of 2020, we are still in a highly emergent healthcare environment in which shifts of circumstance can happen quickly, but here is a present-moment snapshot of what we see happening around us in health systems across the country due to COVID-19:

ORs: Most hospitals canceled or postponed elective surgeries to prepare for the influx of patients from COVID-19. In a poll we did on April 15, 2020, more than 80 percent of US hospitals had seen elective surgery volumes decrease by 70 percent or more since March 15, 2020. This has created a seemingly paradoxical situation in which a great many ORs have been sitting empty, even as there is a popular perception of hospital overwhelm. Given that ORs are the economic backbone of hospitals, a top priority for hospitals going forward will be to address the large backlog of surgeries and restore the elective surgery caseload.

Emergency departments: The stay-at-home orders have had the positive side effect of greatly reducing the number of car accidents and other traumatic injuries requiring ED services, so ED volumes are way down. Consequently, many EDs have furloughed staff.

Inpatient beds: In hard-hit regions such as New York, Seattle, and Florida, most inpatient units are running at very high or overflow utilization rates, while in low-impact regions still waiting for the peak of COVID-19, units are running at very low rates. Many are concerned that as elective surgery and ED volumes return, there will be critical bottlenecks in areas such as the availability of ICU rooms.

Clinics: Case volumes have dried up as stay-at-home orders across the country have kept most patients home. However, telemedicine has picked up a substantial volume of clinic visits, even complex ones. An executive director at a leading US cancer center indicated to us that many top cancer institutions have moved 80 percent of their oncology visits to telemedicine. Some of this shift to telemedicine is likely to endure after the crisis resolves, as both patients and providers have now gained comfort and familiarity with the new platform.

Infusion centers: Volume is also down—between 10 and 25 percent on average, depending on the region of the country (but less than OR volumes). This reduction is due to two main factors. One is that infusion patients are often immunocompromised, so neither they nor their physicians are willing to risk their potential exposure in clinics. The other is that new cases are not being diagnosed because patients are putting off routine screenings and nonemergency clinic visits. Infusion centers, too, will need to address a backlog of patients who will need to get back on their chemotherapy and other infusion regimens after the immediate crisis is over.

Equipment: Ventilators, respirators, face masks, and other items have been in extremely short supply, as we well know.

Every asset is operating under highly abnormal "shock" conditions in regard to demand-supply. However, there *is* a step-by-step process that hospitals can follow to optimize their response, both during the emergency itself and as the system returns to "normal."

Let us illustrate the type of process that can be helpful, using ORs as an example.

RESTORING DEMAND-SUPPLY BALANCE IN THE OR: ELECTIVE SURGERY CASELOAD POST-COVID-19

After a crisis in which demand-supply balance is seriously disrupted, restoring elective surgery volumes requires careful planning and execution. An effective response must include identifying the backlog, managing block versus open time, opening up more time to do more cases, and accommodating the pent-up volume—all within the context of each hospital's resource constraints (rooms, staffing, locations, equipment, bed capacity, etc.).

Here is the seven-step process we are using with our partner institutions during the COVID-19 crisis to restore the elective caseload. As you will see, it entails a combination of analytical tools, process and policy changes, and tight execution.

1. ESTIMATE THE BACKLOG OF CASES AND HOW LONG IT WILL TAKE TO RECOVER

The first step is to get a handle on the demand that has been built up during the crisis. There are online calculators and other resources that perioperative teams can use to estimate the backlog that will build up by the time elective volumes return, such as the one found at http://covid19.iqueue.com.

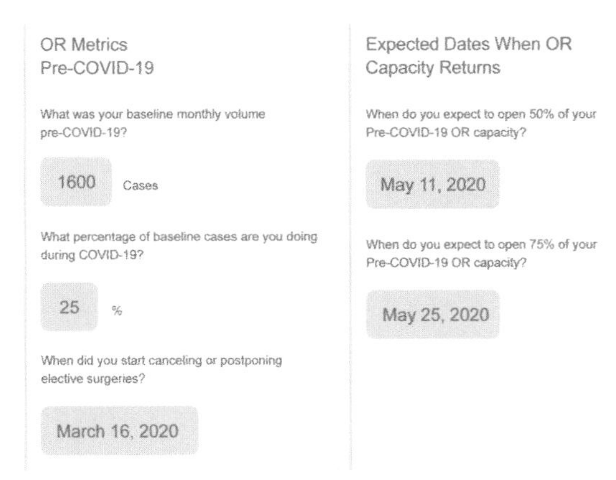

Results

This calculator estimates how long your institution will need to complete the elective surgeries postponed during the COVID-19 crisis. The analysis is based on the size of the backlog during COVID-19 and recovery strategies you can undertake post pandemic.

Based on your inputs, we estimate that:

-You backlog of cases will continue to accumulate until June 15, 2020 and reach the maximum of 2,975 cases.

- After June 15, 2020 you will be able to accommodate 320 additional cases per month, on top of your baseline volume.

- It will take you 9.3 months to recover the backlog, which means it's estimated that you will have cleared the backlog by March 20, 2021.

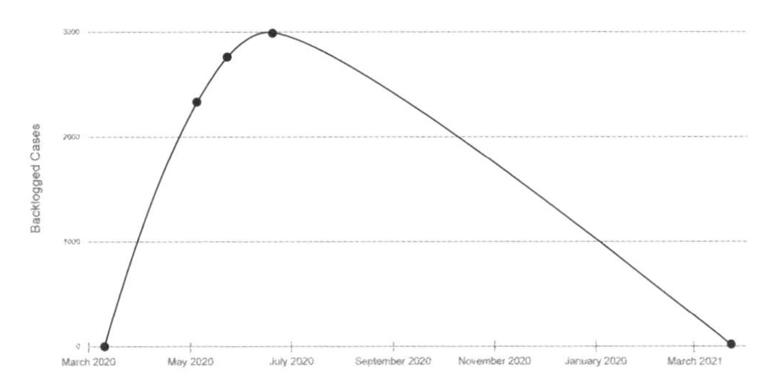

Estimating the backlog built up due to postponing elective surgery volume

This calculator estimates how long the backlog will continue to accumulate and how much time an institution will need to complete the elective surgeries that were postponed during the crisis. The analysis is based on the size of the backlog as well as the postpandemic recovery strategies each hospital is able to undertake (e.g., increasing prime-time utilization and/or extending operating hours, even perhaps working weekends).

Based on the inputs, the calculator estimates the number of additional cases per month the institution will be able to handle on top of its baseline volume and also predicts how many months it will take the institution to clear its surgery case backlog.

2. IDENTIFY REAL SURGICAL CAPACITY AND CONSTRAINTS

Identifying how much surgical capacity can be brought back online, and how fast, depends on understanding each system's "real OR capacity." This is a multifaceted calculation that demands thorough analysis of hospital bed capacity, OR staffing, equipment, and all perioperative support areas (preop, PACU, and SPD).

We polled 425 health systems regarding their biggest constraints in regards to coming fully online after the pandemic, and found the following (respondents could pick multiple options):

- Sixty percent felt their biggest constraint was having enough staffing (nursing and anesthesia) and knowing the flexibility of working hours for their staff.
- Thirty-eight percent identified downstream bed availability as a top constraint.
- Thirty-seven percent said they were concerned surgeons might not cooperate if the block schedule needed to be

changed.

- Thirty-seven percent cited ancillary services (labs, MRI, and others) as a likely bottleneck.
- Eighteen percent said the availability of PPE (personal protective equipment) might be their biggest constraint.

Two tools for surveying staff and estimating available nursing hours can be accessed here:

Web survey for staff preferences and constraints:
https://bit.ly/2YQel6G

Nursing hours calculator:
https://bit.ly/37JXS7U

3. MANAGE SUPPLY OF OR CAPACITY BY RETHINKING THE BLOCK SCHEDULE

During the crisis, surgical departments should be working diligently toward the goal of maximizing OR utilization once the crisis has ebbed. At LeanTaaS, we are cooperating with health systems on this effort, urging them to consider temporarily/partially revising the block schedule and/or substantially increasing the lead time for auto-releasing the majority of blocks. Using a wait list to fairly allocate some of the unblocked time can also be helpful.

Many systems are finding their EHR is not sufficient to capture the size and character of the backlog and are using a manual system or asking clinics or department leads to track the backlog. Here, once again, using online tools with built-in math intelligence—such as a wait list and priority estimation tool—can be a huge help:

My Clinic's Backlog

Priority levels

● P1	● P2	● P3	● P4
Potential fast progression of disease affecting outcome by delay	Severe pain or dysfunction and/or disability but no fast progression	Mild pain or dysfunction and/or disability but no fast progression	No pain and mild dysfunction and/or disability but no fast progression

Surgeon	Position in queue	● P1	● P2	● P3	● P4
Dr. Brandon Flores Patient wait times: 32 days Backlog: 35 cases	4th	3	12	20	0
Dr. Bernard McKinney Patient wait times: 28 days Backlog: 25 cases	10th	0	6	13	2

Enabling clinics to capture the backlog electronically

4. MAKE IT EASY FOR CLINICS TO RELEASE BLOCKS AND TO FIND AND REQUEST THE RIGHT OPEN TIME

Now is the time, as never before, to make a dedicated effort to efficiently match supply with demand and to significantly increase transparency—by adopting sophisticated, analytics-based tools. Upgrading in this manner can help ensure easy block release and easy access to the right OR times for the right clinics, as well as easy requests and easy approvals.

Now is the time, as never before, to make a dedicated effort to efficiently match supply with demand and to significantly increase transparency—by adopting sophisticated, analytics-based tools.

All of this functionality can be easily configured in a customized way to fit each hospital's needs. During "normal" times, this kind of optimization might be seen as a nice extra; it now becomes essential in helping surgical teams work through backlogs as quickly and safely as possible.

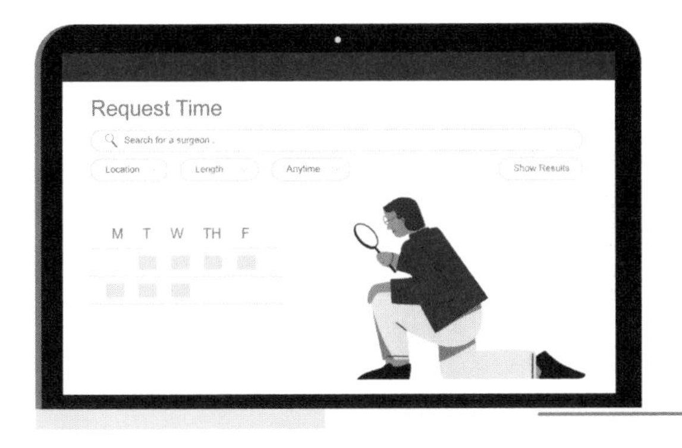

Optimized tools can simplify searching for and requesting open OR time.

5. EXECUTE AS ONE TEAM WITH A COHESIVE PLAN

Quality execution is as important as good strategy. Leveraging every asset to its fullest requires unprecedented teamwork within the peri-operative team and across the hospital. Strong and capable leaders must communicate the urgency of the situation and rally the team around a common goal. It may be necessary to adopt a war-room mentality for a while, using frequent huddles to make decisions based on current constraints and changing daily dynamics.

6. CONTINUALLY MEASURE AND ITERATE

Teams must continually monitor their performance and progress (for iQueue users, this can be done easily through the Analyze module) and iterate on their processes to gain better efficiencies.

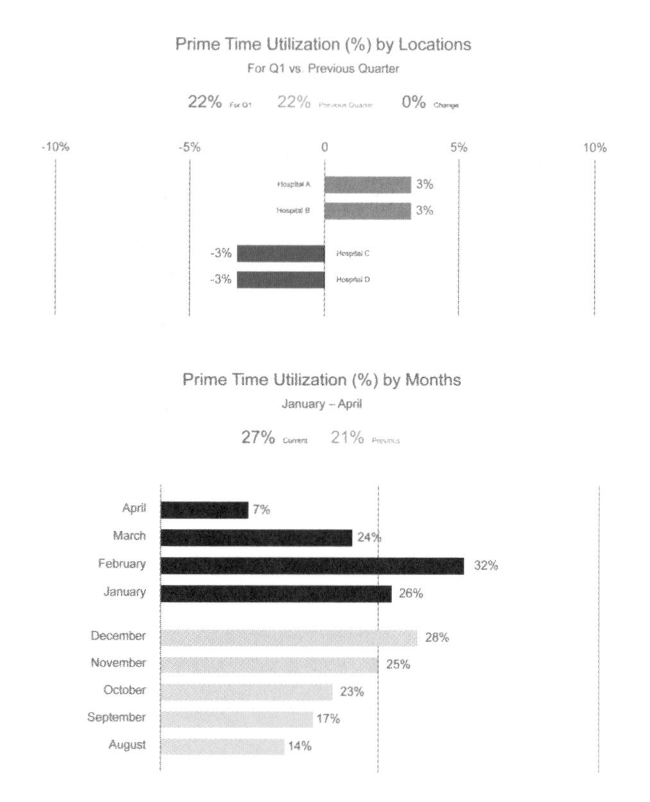

Tools such as the above can provide clear data for improving processes.

It is vital to periodically ask key questions such as the following:

- Do we have a better view of the backlog now?
- How many additional cases did we actually accommodate by extending weekday hours (as compared to the plan)?
- By what percentage did we increase prime-time utilization (as compared to the plan)?
- How is staff morale?

Based on the answers to these questions, the team adjusts/realigns its strategies as needed and reruns the model. Maybe it's time to close some rooms? Reduce some extra shifts?

7. PREPARE FOR THE NEXT WAVE

Teams must also actively prepare for history repeating itself. During the Spanish Flu pandemic, there were three distinct waves of illness, starting in March 1918 and subsiding by summer of 1919. The infections peaked in the United States during the *second* wave—not the first one—in the fall of 1918. This highly fatal second wave was responsible for most of the US deaths attributed to the pandemic.

If COVID-19 comes back in future waves, we may need to go through steps one through six again—and perhaps take additional measures based on the anticipated severity of subsequent waves. This is the hard reality, and we all need to be prepared for it. We must hope for the best and plan for the worst.

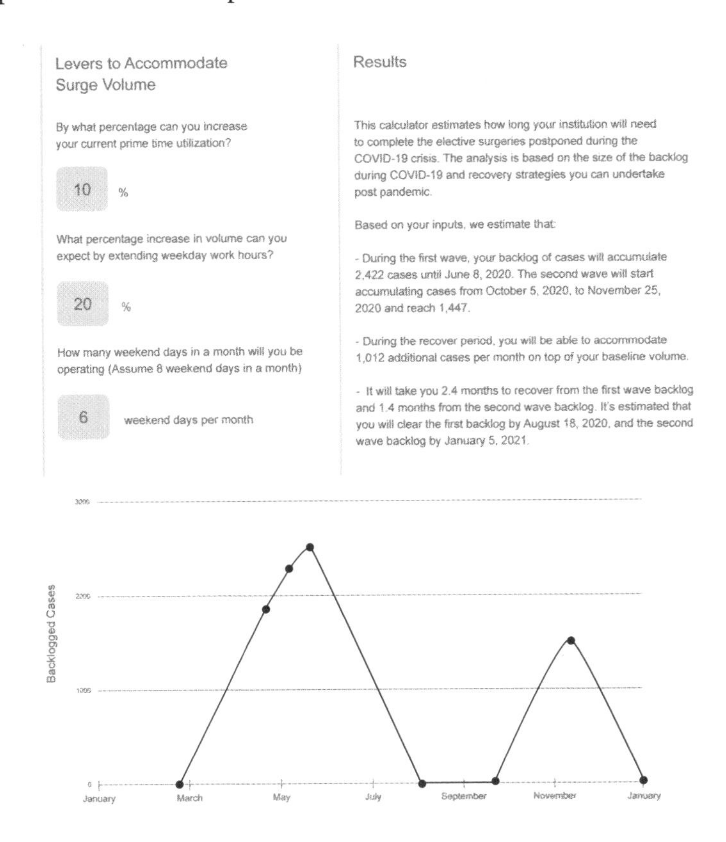

Intelligent tools can help you plan for the next potential crisis.

***

All service areas within healthcare systems need to take intelligent optimization steps, both *as* the crisis is unfolding and later when dealing with backlogs, in order to ensure that all of our precious healthcare assets, human and material, are leveraged to their greatest efficacy. We looked at ORs as an example, above, but similar processes must be followed in other areas—for example, the Penn CHIME has created a model for estimating the need for inpatient bed units that acquired a log of good usage during the early days of planning for bed capacity during the COVID-19 crisis.

As we noted, we are writing these words in the middle of the COVID-19 pandemic, and the situation on the ground is fluid and changing. We are all learning new things on a daily and weekly basis. At LeanTaaS, we are working hard to provide new tools and strategies that can help optimize services in all the healthcare areas covered in the book—ORs, clinics, infusion centers, EDs, inpatient units, and more—as the crisis unfolds and resolves. We look forward to offering even better and more targeted solutions in the future, whether we face another wave of the COVID-19 pandemic or some other new and unforeseen challenge.

When circumstances are at their most challenging, using powerful, math-based, optimized solutions becomes *even more* important, not less so.

ACKNOWLEDGMENTS

We are fortunate and privileged to have written a book about a subject that matters.

Fortunate to be able to work with, and learn from, an incredible team at LeanTaaS. Sincere thanks to all of you. You have already built and deployed many of the tools we have written about here. Without your imagination, dedication, and commitment, this book would have been largely theoretical. Thank you for letting us serve you as leaders in building our amazing company, where we all work hard every day to help our customers deliver Better Healthcare through Math.

Privileged because we can't think of many better missions to embrace—and better ways to make a living—than the pursuit of operational excellence in healthcare.

A very special thanks to Dr. Allan Kirk, chairman of the Department of Surgery in the Duke University School of Medicine and surgeon-in-chief for Duke University Health System, for taking the time to write the foreword.

Thanks also to the dozen or so named individuals and institutions who have made specific contributions to the book and to the hundreds of other individuals at partner institutions who have become friends in our collective journey. You have pushed us and shaped our thinking. As a result, this book has been able to offer practical solutions that *actually work* despite the inherent challenges of trying to change operational processes in the complex world of healthcare. Thank you to all our customers for your foresight and your desire to innovate with us, not just "admire the problem."

Profound gratitude to our respective parents, who gave us the love of learning and the curiosity to solve hard problems; to our children, who are a daily reminder that working to improve healthcare operations is a noble pursuit despite its difficulties; and to our families for being 100 percent behind us and supportive of our writing the book.

Finally, thank you to those who helped us create the book—to ForbesBooks and Advantage for approaching us with the idea and helping us actually write the book. It has been a hard process and one we might never have embarked on without your "push" and the support and resources that enabled us to get here.

Together we aspire to do much more.

—Sanjeev and Mohan